ETHICAL DECISION MAKING IN PHYSICAL ACTIVITY RESEARCH

John N. Drowatzky, JD, EdD
The University of Toledo
Toledo, Ohio

Human Kinetics

Library of Congress Cataloging-in-Publication Data

Drowatzky, John N.
Ethical decision making in physical activity research / John N. Drowatzky.
p. cm.
Includes bibliographical references and index.
ISBN 0-87322-473-6
1. Sports medicine--Research--Moral and ethical aspects.
2. Kinesiology--Research--Moral and ethical aspects. 3. Exercise--Research--Moral and ethical aspects. I. Title.
RC1235.D76 1996
174'.28--dc20 95-42649
CIP

ISBN: 0-87322-473-6

Developmental Editor: Marni Basic; **Assistant Editors**: Susan Moore, Ed Giles, and Lynn Hooper; **Editorial Assistant**: Jennifer J. Hemphill; **Copyeditor**: Elaine Otto; **Proofreader**: Pam Johnson; **Production Manager**: Kris Ding; **Typesetting and Text Layout**: Sandra Meier; **Text Designer**: Judy Henderson; **Cover Designer**: Keith Blomberg; **Illustrator**: Jennifer Delmotte; **Printer**: Versa Press

Printed in the United States of America

10 9 8 7 6 5 4 3 2 1

Human Kinetics
P.O. Box 5076, Champaign, IL 61825-5076
1-800-747-4457

Canada: Human Kinetics, Box 24040, Windsor, ON N8Y 4Y9
1-800-465-7301 (in Canada only)

Europe: Human Kinetics, P.O. Box IW14, Leeds LS16 6TR, United Kingdom
(44) 1132 781708

Australia: Human Kinetics, 2 Ingrid Street, Clapham 5062, South Australia
(08) 371 3755

New Zealand: Human Kinetics, P.O. Box 105-231, Auckland 1
(09) 523 3462

Dedicated to the memory of

Minnie Louise Drowatzky,
my first teacher

CONTENTS

PREFACE

Most researchers active today learned that to be successful they needed motivation, skill in using the scientific method, and knowledge of the literature regarding the problem in question. Their professors and textbooks taught them to seek research problems by considering whether the problem was of interest, the research feasible, their training appropriate, and whether the problem would make a significant contribution. In short, potential researchers were taught that research is problem solving used to understand humans, animals, and the environment. Few graduate students were taught that researchers had an obligation to consider potential ethical considerations when developing their research problems or how to analyze any ethical issues that might arise. Thus, the typical researcher thought little about ethical issues that might be present in research.

When federal agencies began publishing requirements for receiving their grants, researchers often regarded the regulations as one more hoop to jump through before obtaining money for their projects. These agency regulations arose, however, as a consequence of the crimes trials after World War II atrocities, when the unethical research practices followed by Nazi scientists were discovered. These war crimes made clear the need to train scientists to recognize ethical considerations in research.

My purpose in this book is to identify the relationship between ethical considerations and the research process. An historical overview of the development of ethical codes relating to research leads to an analysis of ethical considerations in the formation, conduct, analysis, and presentation of the research problem. To assist with the process of identifying ethical issues that arise in research, examples are provided throughout the text. These examples, while appropriate to research in general, are taken mainly from the areas of exercise science, physical education, and sport science. Other examples come from cases that were highly publicized in the news media. While the specific cases are taken from sport-related fields, the issues raised are universal.

This book will look at ethics and integrity in science and research using the experimental process as the framework for evaluation. Ethical questions arise at all stages in research. *Ethical Decision Making in Physical Activity Research* focuses on ethical concerns that can arise during

- the selection and formation of the research question;
- the work of institutional review boards;
- methodology, including the use of human and animal subjects, data collection, and statistical analysis;
- the presentation of results, including publication; and
- practices currently used to prevent fraud and misconduct in research (e.g. institutional responsibility, individual responsibility, and publication practices).

A series of case studies have been included to help develop sensitivity toward ethical issues in research and to assist in the development of ethical thinking. Readers can use the case studies to determine whether ethical dilemmas are present in certain situations, to identify what those issues are, and to consider the concerns such dilemmas present. Next, the behavioral options present can be determined, and finally, the rightness or wrongness of the choices can be determined. When determining the rightness or wrongness of actions, the following questions should be answered: Why is the choice right or wrong? If the action is wrong, what needs to be changed to make it right? In closing, suggestions for actions to minimize or prevent fraud will be presented.

You will find the issue of ethical practices in research presented both in the text narrative and in questions, in addition to case studies and case problems presented throughout the book. These cases are based on situations commonly encountered by practicing professionals, some of them upon actual ethical problems that have occurred. I encourage you to identify the ethical issue, if present, and to decide the appropriate course of action to resolve it.

In the appendices you will find the regulations relevant to research using human subjects and animal subjects. Researchers today must know these regulations if they actively use human or animal subjects.

Ethical Decision Making in Physical Activity Research will be useful as a supplemental text for classes in research methods or for seminars dealing with ethical practices in research. The intended audience includes professionals and students in exercise science, kinesiology, and sport science who have an interest in conducting and evaluating research. Students and scientists pursuing research in these fields face the ethical problems described in this text and face questions similar to those raised here.

I wish to acknowledge the contributions of my students, whose questions led to the development of this book. Teaching has always been a learning experience for me. My students and I explored issues that were often ignored in standard research textbooks.

CHAPTER 1

THE NATURE AND ROLE OF ETHICS

Creativity and the pursuit of knowledge have been human goals since time immemorial. Genesis emphasizes this in the story of Adam and Eve and their involvement with the tree of knowledge. Throughout recorded history men and women have attempted to understand and explain phenomena they observed. These phenomena range from natural events to cultural differences. Common sense has proved not to be a good provider of knowledge as it lacks any means for the evaluation of knowledge and does little to produce new knowledge.

Likewise, tenacity, authority, and intuition have been insufficient methods of obtaining knowledge. *Tenacity* refers to the proposition that one holds to the truth because it is known to be true and it has always been known to be true. Here individuals tenaciously hold to their beliefs, tenacity tending to strengthen these beliefs even if doubt is cast on their validity. Knowing by way of *authority* means relying on established beliefs. Information is accepted as true because it is published or imparted by an authority figure. Using the intuitive, or *a priori*, method of knowing, propositions are accepted when they are "agreeable to reason" or are self-evident. Thus, with this method of knowing the premise is that if something "stands to reason," it is self-evident and true: humans can reach the truth because their natural inclinations tend toward truth (Kerlinger 1966).

Science, on the other hand, provides a systematic and controlled procedure for evaluating ideas. The scientist acquires knowledge through research involving the use of the scientific method. This method sets forth a form of inquiry appropriate to all fields of study, its principles as applicable to research in history, English, or philosophy as they are to research in physiology, psychology, physics, or chemistry. Scientific research, according to Kerlinger (1966, 13), "is systematic, controlled, empirical, and critical investigation of hypothetical propositions about the presumed relations among natural phenomena."

The scientific method was developed to provide a necessary element for obtaining information: self-correction through the use of a built-in series of checks. Over the years, the scientific method has evolved to enhance our ability to explain our universe and ourselves. The scientific method blends logic and practical experience and provides for the self-correction of errors. This self-correcting aspect involves the empirical testing of the phenomena we are studying. Here the researcher puts personal belief to a test. Subjective belief is checked against objective reality. The scientific method provides a systematic, ordered, and controlled framework through which this check can occur. The basic steps employed by the scientific method are

- **identification of the problem**—when experiencing an obstacle to understanding, express the problem in a reasonably manageable form;
- **hypothesis**—develop a conjectural statement or tentative proposition about the problem;
- **reasoning**—deduce the consequences of the hypothesis formulated and, if necessary, revise the problem or abandon the study; and
- **observation-test-experiment**—put the problem to an empirical test; examine the hypothesis or the deduced implications of the hypothesis.

Empirical study means that the phenomenon being tested is subjected to experience and observation and can be replicated by others.

When used appropriately, the checks are anchored as much as possible in authenticity existing outside the scientist. All statements, even if they appear true, will be tested for veracity. If the hypothesis appears to be supported by the experiment, alternative hypotheses will also be tested to see if they cast doubt on the first hypothesis. The experimental and testing procedures are open and published so that others can attempt to replicate the experiment. The whole process was developed to provide objectivity in research.

Any process can be abused, and the scientific method is no exception. To make the scientific method effective, faithful adherence to its procedures is required so that the researcher's personal biases, values, attitudes, and emotions are not factors. Thus, while the scientific method provides the way to make an unbiased evaluation of knowledge, the characteristics of the investigator remain important. The following articles from the Toledo *Blade*, 30 April 1994, and *USA TODAY*, 2 May 1994, demonstrate recent abuse of the scientific method.

Doctor Quizzed in Faulty Study

PITTSBURGH—Federal officials have asked the University of Pittsburgh to determine whether Dr. Bernard Fisher did anything wrong while supervising a $17 million breast cancer study that was tainted by fraudulent data. The study was based on research that included falsified data from St. Luc Hospital in Montreal.

Federal officials want to know whether Dr. Fisher, who resigned March 28 [1994] as director of the national Surgical Adjuvant Breast and Bowel Project, and his chief statistician, Dr. Carol Redmond, included data they knew to be false when they submitted their research.

More Faulty Data in Breast Cancer Study

Government auditors are turning up more problems in a landmark breast cancer study.

Teams are reviewing some 1,800 patient records used in the study that concluded lumpectomies are effective alternatives to breast removal for breast cancer patients. The probe began in March [1994] after officials discovered fraud by a Canadian researcher.

The *Chicago Tribune* reported Sunday that the Memorial Cancer Research Foundation of Southern California, Beverly Hills, Calif., submitted data

- on women who were ineligible for the study;
- on a woman described as ''alive and well'' who, in fact, had died; and
- about women who had not consented to being in the study.

''It's a mess right now,'' says Dr. Bruce Chabner, a top National Cancer Institute official. Chabner is troubled because study leaders knew about the California data errors in 1990, yet didn't tell the NCI. They also failed to report fraud by a Montreal doctor involved in the same study. ''They didn't want us to know about things that were a problem,'' Chabner says.

Several newer studies also support lumpectomy, NCI officials say, and this study was large enough so that a few faulty data points won't alter its conclusion. But the new problems ''raise questions about the database itself,'' Chabner says.

NCI suspended the study group, coordinated by the University of Pittsburgh, and ousted its leader, Dr. Bernard Fisher, in March. Larry Weinberg, a spokesman for the California researchers, acknowledges their data may include ''some anomalies and relatively inconsequential errors.''

As shown here, dishonest and unethical researchers can circumvent the checks and balances built into the scientific method. As the press has illuminated more cases of fraudulent and dishonest research that has obtained government funding, pressure has increased to reduce the possibility for such occurrences. Ethical issues in science have been identified and ways to prevent dishonesty have been proposed. The application of ethical systems designed to maintain integrity in science and research has been a result of this process.

Ethics

A number of disciplines analyze human behavior. Sociologists study the development, structure, interaction, and collective behavior of organized groups. Psychologists explore human behavior from developmental, motivational, and learning perspectives. Anthropologists study human beings in relation to distribution, origin, classification, and relationship of races, physical character, environmental and social relations, and culture. Some researchers attempt to describe behavior, while others look for cause-and-effect relationships. In each case, these scientists tell what behavior is or was.

Ethics views human behavior from a perspective of good or bad. Ethical statements are developed to prescribe behavior as directives that tell us what we ought to do in the situations we encounter. A goal of ethics is to develop a system of general ethical statements that provide a distinct universal prescription (one that applies to everyone) detailing which behaviors are right and which are wrong.

According to Reagan (1971), people often define ethics as what they feel they should do. However, ethics is not concerned with what people think or feel they ought to do, but with what they really ought to do. Interpersonal conduct, which is the focus of ethical study, differs from legally mandated behavior or custom. Each approach deals differently with the behavior in question. Law and morals are similar as they deal with the more important human behaviors, whereas custom or etiquette deals with minor concerns. For example, both law and ethical rules may agree that a particular behavior, such as murder, is wrong. In other cases, they may disagree, as in instances where a particular behavior, such as abortion, may be legal yet may also be viewed as morally wrong. Thus, ethics and law govern behavior from different perspectives. Law looks at the preservation of civil order and individual rights, while ethics is concerned with morally correct action. In contrast, custom views behavior in terms of socially correct actions.

Any discussion of ethical behavior requires a common ground, based on a uniform understanding and usage of words. Therefore, the following definitions from *Webster's New International Dictionary of the English Language* are used here.

- **ethical**—of or relating to moral action, motive, or character; containing precepts of morality, moral feelings, duties, or conduct.
- **ethics**—moral principles, quality, or practice; a system of moral principles; the morals of individual action or practice.
- **moral**—manner, custom, habit, way of life, conduct characterized by practical excellence, or springing from, or pertaining to, man's natural sense of what is right and proper. Of or pertaining to morals; designating, or relating to, the science or philosophy of conduct; hence, relating to, or regarded with respect to, the qualities and considerations with which morals deal, as questions of right and wrong or virtue and vice. Conforming to, or embodying righteous or just conduct, or the dictates of the moral sense. Relating to virtuous conduct or natural excellence as distinguished from civic or legal righteousness.
- **integrity**—state or quality of being complete, undivided, or unbroken; wholeness; entirety. Moral soundness; honesty; freedom from corrupting influence or practices.

- **honesty**—characterized by integrity or fairness and straightforwardness in conduct, thought, speech. Upright; just; equitable; trustworthy; truthful; sincere; free from fraud, guile, or duplicity; not false.
- **right**—being in accordance with what is just, good, or proper; agreeable to a standard; conforming to facts or truth.
- **wrong**—an injurious, unfair, or unjust act; something wrong, immoral, or unethical; principles, practices, or conduct contrary to justice, goodness, equity, or law.
- **fraud**—quality of being deceitful; deception deliberately practiced with a view to gaining an unlawful or unfair advantage; deceitfulness; deceit; trickery.

These definitions are circular and interrelated, and they are difficult to quantify. Attempts to justify moral principles by definitions would be difficult, if not impossible, whenever there is substantial disagreement over the definition of the term being used. Common themes included when defining *ethics*, *moral*, *integrity*, and *honesty* are ''right and proper, honesty, free from fraud and deception, and freedom from corrupting influence or practices.'' *Fraud* and *wrong* have the opposite meanings, namely, ''unfair, unjust, immoral, deceitful and trickery.'' Even with agreement on definitions, questions remain. For example, is one right better than another right? Is one wrong worse than another? Can we rank order values? Is there a difference between morality and ethical behavior? Is fraud the absence of moral behavior, or is fraud an entirely different type of behavior?

If we are to determine the right or wrong of a behavior, person, or thing, what standards do we use to judge value? Do we attempt to limit ourselves to nonmoral value of the sort that we place on cars, tennis racquets, food, and other material objects? We regularly assign these things values of good, bad, or average without ever using a moral approach. We compare one item with another and specify a value rating based on how well it accomplishes its purpose. We also use instrumental value to determine good or bad, rating something on how well it accomplishes something else. Whether the ''newest'' fashion is good or bad depends on how well it satisfies those who chose to purchase it. When a thing is described as good in itself, such as happiness or peace of mind, it has intrinsic value. When evaluating people, we often describe their character traits as good or bad. Since we can't directly measure character traits, we rely on behavior to judge a person as good or bad. While it is desirable to develop good character traits, such an approach does not answer all ethical questions. Values are relative, and questions remain. For example, is a person immoral for failing to do something that was beyond his or her ability to do? Is a person moral for doing something that he or she must do?

A major problem in the development of ethical principles is that they cannot be verified, as scientists like to do with their data. However, they can be confirmed or disconfirmed relative to the evidence. A potential problem exists with this process because, if a researcher selects evidence carefully, then almost any empirical statement can be verified relative to the evidence used. It is therefore necessary to evaluate empirical statements relative to all the available evidence rather than to engage in a process of selecting which evidence to use and which to ignore. No empirical statement can be evaluated in isolation. Rather, it must be evaluated as a part of the theory in which it plays a part. Consequently, empirical statements are indirectly evaluated while the theories are directly evaluated.

Systems of Ethics

Several approaches to ethics have been developed over the years. Not only do these approaches come from different perspectives but they involve different subject matter. *Normative ethics* is the study of rules, norms, or criteria by which we judge actions as right or wrong. Normative ethics therefore is a rational attempt to determine how one should behave in a particular situation. A second subject matter is denoted by *metaethics*, which is a study of what we mean when we make certain statements. For example, metaethics attempts to answer questions such as, "What do we mean when we use *good* as an ethical term?" *Ethical theories* are developed to establish a coherent relationship between the different elements of the theory and to justify particular moral judgments. *Ethical rules and principles* are used to build a theory. A well-developed theory will enable us to show whether our moral judgments are correct and reasonable in light of the theory's rules and principles. To illustrate how a system of ethics will function, two systems—situational ethics and rule utilitarianism—are briefly presented as examples.

Situational Ethics

Some people will argue that no general ethical rules are applicable, as having general rules that apply to all individuals in all situations is impossible. Each person is unique. Therefore, there is no universal human nature. Since we are unique, each situation in which we find ourselves is also unique. Because of this, we cannot use class logic because when we do so, individuals are treated as though they are not unique. Proponents of this position follow the theory of situational ethics, which requires that each action be evaluated on its own merits and without any reference to general rules or obligations. In other words, there are no universal ethical rules that must be followed. Situational ethics follows the premise that you should evaluate each situation or case and do what you think is right in that situation. This ethical practice means that there is no one correct answer or behavior, as it always depends on the unique situation at hand. Returning to the example of scientific misconduct, actions based on situational ethics will often lead to results such as were present in the news item that follows. This is a follow-up article, published May 19, 1994, in the Toledo, Ohio, *Blade*. It deals with the same incident reported earlier in this chapter.

Motive Was Patients, Says Doc Who Faked Breast-Surgery Study

BOSTON (AP)—A surgeon who faked records in a landmark breast cancer study said he only wanted the best treatment for patients and didn't believe the project's rules had to be "followed blindly."

Dr. Roger Poisson of St. Luc Hospital in Montreal was one of the chief contributors to the study which changed the way early breast cancer is treated. His falsifications of dates in a handful of cases cast doubt on the study, calling into question the now standard practice of treating breast tumors by removing the cancerous lump rather than the whole breast.

Experts have since assured patients that his falsifications did not change the study's overall results, and other research confirms that lumpectomy is acceptable treatment. Nonetheless, the affair triggered a furor, in part because of a delay in making the fraud public. Nearly four years elapsed from when officials first became suspicious of Dr. Poisson's data and the disclosure in an article in the *Chicago Tribune* two months ago.

Dr. Poisson's explanation was one of several letters about the affair published in today's *New England Journal of Medicine*. Among the other versions of events was a step-by-step account by Dr. Bernard Fisher of the University of Pittsburgh, the study's coordinator, who was recently dismissed from the project because of the uproar. Dr. Fisher said that by the time a two-year federal investigation was over last year, it was clear that the fraud had no impact on the study's results, so there was "no issue of public health."

"We failed to recognize that the public might interpret that lack of an immediate announcement or publication of a reanalysis as a sign that we were less than forthright or even that we had concealed information that could have affected the treatment of women with breast cancer," he wrote.

The lumpectomy study, published in 1985 in the *New England Journal of Medicine*, was part of the National Surgical Adjuvant Breast and Bowel Project, which Dr. Fisher headed. In the study, women with breast cancer were randomly assigned to have lumpectomies or mastectomies. All participants were supposed to meet explicit rules regarding their health and treatment. The goal was to ensure they were as alike as possible so valid comparisons could be drawn.

Of 1,511 women in the study, 354 were Dr. Poisson's patients. The U.S. Office of Research Integrity concluded that Dr. Poisson faked records of six of them. Each of the six should have been excluded from the study because too much time had elapsed since biopsies were performed to diagnose their cancers. Dr. Poisson changed the date of their biopsies so they could take part. He said he believed study participants would get better therapy and follow-up than had they not been enrolled.

> "I felt that the rules were meant to be understood as guidelines and not necessarily followed blindly," Dr. Poisson wrote. "For me, it was difficult to tell a woman with breast cancer that she was ineligible to receive the best available treatment because she did not meet one criterion out of 22, when I knew this criterion had little or no intrinsic oncologic importance," Dr. Poisson wrote. His letter didn't mention the health status of the six women.
>
> In an editorial, the journal's two top editors questioned whether Dr. Poisson's motives were entirely selfless, since his hospital was paid according to the number of patients in the study.

If we evaluate the statements and decisions made by Drs. Poisson and Fisher quoted above, several questions about professional ethics arise. Was Dr. Poisson engaging in situational ethics behavior when he said that "rules were meant to be understood as guidelines and not necessarily followed blindly?" Does it make any difference to your answer that Dr. Poisson's hospital received financial benefit based on the number of subjects participating in the study? For women who did not qualify as subjects for the study, was it ethical that they would not "receive the best available treatment" if such was indeed the case? Was it ethical to break the scientific protocol arbitrarily and taint the research that would be of benefit to many other women by including ineligible women as subjects even if it were true that nonsubjects did not "receive the best available treatment"? In other words, do you sacrifice the good of many for the good of a few? Was it ethical that Dr. Poisson made the unilateral decision that "this criterion had little or no intrinsic oncologic importance"? What ethics were involved with Dr. Fisher's position that "the fraud had no impact on the study's results, so there was 'no issue of public health,' " which led to his failure to blow the whistle on Dr. Poisson's fraudulent data? Was it right for him to withhold this information unilaterally? Are such researchers above the rules published by the granting agencies and accepted professional standards of conduct that govern their ability to make such decisions?

When dealing with the issue of universality in ethics as well as interpretation of research data, it's important to keep in mind that every scientific theory is an attempt to generalize; in fact, that's why the sample used in a study must be representative of the population from which it comes. Without the ability to generalize, chaos would result. This does not mean that we should ignore individual differences. Rather, humans are similar enough that we can use generalizations while being alert to the uniqueness of individuals involved. We can continue to apply our theories and practice, keeping in mind when we make predictions that our theories apply to groups of people, not individuals. The adoption of situational ethics would not give predictability and universality to ethical conduct. There would be no universal right or wrong acts.

Rule Utilitarianism

Utilitarianism is another approach to ethical conduct. It resembles situational ethics, as all actions are evaluated by their consequences (i.e., whether they produce greater good than evil in the world). Like situational ethics, only a particular action performed by a particular person in a particular situation is evaluated. However, several basic problems have led to the discarding of this approach and the development of rule utilitarianism.

Rule utilitarianism is a philosophical approach developed by social philosophers who were disappointed with the potential injustices that resulted from the utilitarian approach. According to Grassian (1992), the utilitarian approach attempted to separate moral foundations from a consideration of universal consequences and psychological and sociological knowledge. Instead, it sought to determine the rightness of an action by its individual consequences. In contrast, those who accept rule utilitarian philosophical approaches will base their ultimate concern on the presence of beneficial consequences, asking the question, "What would be the consequence if everyone were to do this?" The approach asking this fundamental question is known as *general rule utilitarianism*, and it makes the moral rightness of an action dependent on the universal consequences of a behavior rather than on independently specified moral rules.

Rule utilitarian positions are further divided into two camps: actual and ideal. Actual rule utilitarians focus on the need for actions to be based on the actual and recognized rules of the society if the behaviors are to have moral justification. Ideal rule utilitarians approach moral justification on the premise that acts are moral if they are justified by moral rules that would be ideal for the society. Hence, the first approach looks to actual rules that are present, whereas the other approach looks to idealistic rules that should guide society.

In practice, many rules and regulations are based on the assumption of actual rule utilitarianism, which postulates that the greatest good for individuals will be to act on the basis of specific moral rules. The use of specific moral rules provides predictability and guidance that would be lost if our specific actions had to be justified by their utility. Utility would leave moral justification to the beliefs of individuals and the society without any uniform moral code, which in turn could lead to chaos and disintegration of society. Thus, according to actual rule utilitarianism, moral life requires that individuals commit themselves to established ways of doing things as defined by rules that define behavior. Perhaps the best example of this is the use of criminal laws that define improper behaviors and establish punishments for engaging in such behaviors. The statements of ethical practice and acts within professions are also examples of the use of moral rules that define appropriate behavior. Given human imperfections, such as the lack of perfect omniscience and concern for others, actual rule utilitarianism accepts moral rules as necessary guidelines that will maximize moral behavior and utility for a society.

Within a legalistic framework, precedence is used to establish ways of doing things, and judges are given some latitude to use their discretion, although they are bound to follow the accepted rules of law. The use of judicial discretion can be seen in the sentencing phase of a criminal trial when the judge weighs factors such as family ties and past behavior to determine the severity of a sentence. In its *Model Penal Code*, the American Law Institute (1958) also accepts the principle of utility when it states, ''Conduct which the actor believes to be necessary to avoid an evil to himself or to another is justifiable, provided that the evil sought to be avoided by such conduct is greater than that sought to be prevented by the law defining the offense charges.''

The purpose of a legal system is to fashion rules that function as constraints on human behavior, that is, indicating acts that society prohibits. According to utilitarianism, these legal rules differ from moral rules that individuals should follow when deciding how to act morally in any given situation. In the utilitarian framework, the moral rules would lead to behaviors based on answers to such questions as ''Is it right to break a moral or legal rule when reasonably assured that such action would have the best consequences?'' rather than engaging in mechanistic and legalistic behavior based on whether the rules permit the act. Although this approach has generally been adopted by society to govern criminal conduct, will the utilitarian framework work in governing professional activities? If so, who decides what are the ''best consequences'' for a profession?

Ethical Problems in Research

Research refers to a class of activities designed to develop or contribute to generalizable knowledge. Generalizable knowledge means theories, principles, or relationships (or the accumulation of data on which they may be based) that can be corroborated by accepted scientific observation and inference. The research process extends from the germination of an idea to the publication of the conclusions. Every scientist publishing a report of completed research knows that other scientists will be reading and evaluating the work. This knowledge alone causes the writer to carefully consider what is written. Practice or clinical action differs from research and refers to a class of activities designed solely to enhance the well-being of an individual patient or client. Therefore, the purpose of practice or clinical treatment is to provide diagnosis, preventive treatment, or therapy (Levine 1981).

Research is generally considered to take one of two forms: experimental or clinical. *Experimental research* is usually conducted in some sort of laboratory and emphasizes the control of all experimental factors. *Clinical research* is usually conducted in medicine and therapies where the treatment of individual patients is followed and then compared with the advantages and disadvantages of other treatments. Such research is not amenable to the controls exercised in experimental or laboratory research. Experimental studies may be classified as either ''pure'' or ''applied.'' Whichever category the research falls into, it relies upon formal, diligent, and systematic inquiry designed to discover or revise facts, theories, and applications. Research is problem solving using the scientific method to guide the inquiry. The commonly used distinction, although many researchers deny any such difference exists, is that ''pure'' science or research does not depend on the examination of practical, concrete problems or issues, whereas "applied" research is atheoretical and attempts to use knowledge to solve existing problems.

Basic researchers usually view themselves as value-free scientists, active in discovering the truth but unconcerned about society's use of their findings. They view knowledge as ethically neutral and their work as objective and morally neutral. This position of scientific nonresponsibility held by "pure" researchers is emphasized when they describe their experiments as "pure" or "basic" research. In contrast, "applied" researchers have been described as engaging in action-oriented research, albeit with a theoretical orientation. Often "applied" research will focus on problems that arise in individuals' adjustment to their society, and such focus comprises much of the research conducted in the social sciences. The primary goal of an "action-oriented" study is to accumulate facts and principles that can be applied to resolve current social problems (Kimmel 1988).

Regardless of one's approach to research, basic or applied, the need for ethical decision making exists. Coupled with the need for ethical decision making is the need to recognize the potential ethical problems that exist in the scientific process. From accounts in newspapers and professional journals, dishonesty and the use of measures that circumvent the scientific method continue in the research community. Newspapers reported researchers manufacturing data at Harvard University and overreporting expenses to increase grant income at Stanford University during the early 1990s. As early as 1980, Brackbill and Hellegers (1980, 20) expressed concern over the ethical conduct of scientists:

> Most scientists are under great pressure to conduct research and publish it. Publication is the sole route to professional success, to salary increases, to tenure, to promotion. Scientists, therefore, regard the terms and conditions of publication as matters of considerable importance. There is no question that ethical review as a gate to publication is an effective means of maintaining ethical standards in research. It is also the most feasible method.

Scientific fraud is a crucial issue that generates more questions than answers. Wrather (1987) indicated that the incidence of fraud in science appears to have increased in recent years. Misrepresentation of results threatens the conduct of scientific research. Important first steps in reducing fraud and misconduct have included the guarantee of confidentiality for whistle blowers and the development of guidelines for use by universities and funding agencies.

The desire for academic success and job security can lead to dishonesty in science and research. Fraud and misconduct in scientific pursuits must be addressed in such a manner as to reduce or eliminate their occurrence. If society is to benefit from scientific research, its belief in the veracity of a study and the information disseminated from it is crucial. Case 1.1 illustrates a practice all too common among professionals these days.

CASE 1.1

A RESPONSE TO PUBLISH OR PERISH PRESSURES

Dr. May I. Cozen partially meets the scholarly activities requirement of her job by presenting several research projects at national conferences each year. She selects conferences that are unlikely to draw the same participants and then submits her abstracts for acceptance on their programs. Dr. Cozen uses the same data in each presentation, but she chooses various titles for the presentations and makes other slight cosmetic changes. She may reword the introduction, list different studies in the literature review section, and present particular conclusions in one presentation and other conclusions in another presentation. Regardless of the changes she makes, however, the data and analysis remain the same. This technique has allowed Dr. Cozen to make three or four presentations in a year using the same data.

Some professionals will readily sign statements that a paper or manuscript has not previously been presented or published. They might argue that the paper hasn't been presented before because it has a new introduction, review, and set of conclusions; only the data—in a different interpretation—have been presented. Consequently, the writers contend, it is a new paper. Is this ethically correct? How much of the paper must be changed before it qualifies as new? Must the data be entirely new, or can a new analysis of data presented in the past qualify as a new paper?

Shore (1991) attempted to determine what types of dishonesty exist in research by analyzing cases from files at the National Institutes of Health (NIH) (using the Freedom of Information Act) and the Harvard University Medical School (see table 1.1). Of the thirty-eight cases reporting dishonesty in laboratory research, twenty-five involved fabrication and falsification, plagiarism, and nonpublication of data that would refute prior research. These cases typified fraudulent and dishonest situations where guidelines and ethical standards are least likely to prevent dishonesty. The remaining thirteen cases involved behaviors in laboratory research that were likely to be influenced by guidelines, including the need for refined data-gathering, storage, and retention procedures; better writing techniques; and proper publication practices. In clinical studies, the problem areas were disregard for protocol (e.g., inconsistent logging of information), poor choice of controls, failure to use blinding codes, and inaccurate logging of data.

Russell (1991) reported similar findings of common transgressions in NIH-funded research. NIH identified failure to retain primary data, overinterpretation

Table 1.1 Problem Areas Identified in Scientific Research

PROBLEM AREA	CASES
Laboratory research	
Fabrication and falsification	16
Plagiarism	7
Nonpublication of data	2
Data-gathering procedures and data storage and retention	9
Authorship	2
Publication practices	2
	38
Clinical research	
Disregard of protocol	6
Logging information	2
Use of controls	2
Use of blinds	2
	12

Data from Shore 1991.

or misrepresentation of results, selective inclusion or exclusion of information, alteration of findings resulting in either no change or a favorable change, plagiarism, and data manufactured without supporting evidence. As was the case with the misconduct findings reported by Shore, some of these transgressions were the result of the researcher's character and cannot be prevented by the adoption of codes of ethical conduct.

These two reports looking at a limited number of studies showed that the checks and balances included in the scientific method of inquiry have frequently been circumvented by fraudulent acts of researchers. Some dishonesty was associated with deliberate misconduct, while other problems resulted from poor organization or techniques. When dishonesty results from fabrication and falsification of data, plagiarism, and selective publication of results, the character of the researcher produces the lack of integrity in science. No ethical codes, policies, or guidelines will totally prevent such behavior from dishonest scientists. Other problem areas identified, such as poor organization, improper techniques, and failure to adhere to protocol, can be improved through codes, policies, and guidelines as well as better mentoring. The Public Health Service and National Institutes of Health felt the problem was extensive enough to warrant the development of definitions of misconduct, which are included in their regulations:

- **"Misconduct"** or **"misconduct in science"** means fabrication, falsification, plagiarism, or other practices that seriously deviate from those that are commonly accepted within the scientific community for proposing, conducting, or reporting research. It does not include honest error or honest differences in interpretations or judgments of data (42 C.F.R. Chapter 1, § 50.102).
- **"Misconduct"** means (1) fabrication, falsification, plagiarism, or other serious deviation from accepted practices in proposing, carrying out, or reporting results from research, (2) material failure to comply with federal requirements for protection of researchers, human subjects, or the public or for ensuring the welfare of laboratory animals, or (3) failure to meet other material legal requirements governing research (NIH 1990).

In any event, the findings reported by Shore and Russell emphasized the need for ethical researchers if scientific integrity is to be achieved and maintained. What does this research indicate? Is there a problem that will be solved by developing and adopting a code of ethics? According to Callahan (1982, 336), "There is a good historical rule of thumb to use in examining any calls for codes of ethics. Very simply, it is that such calls normally seem to arise when a profession, field, or discipline is in an internal state of disarray. . . . They arise as a response to internal tensions and external pressure."

Most of the problems described above relate to dishonesty and/or fraud on the part of scientists. Ethical problems, rather than dishonest conduct, can also appear from the issues involved. In such cases, sensitivity to potential ethical issues is necessary for the decision-making process to be effective. However, sensitivity alone will not be sufficient for resolving ethical questions that may pertain to the body of knowledge under investigation, or the conduct of the research, in such manner as to protect the rights of participants and society.

Studies such as those cited above have shown that ethical questions arise from several factors. First, most ethical problems arise from situations where conflicting values also exist. If no difference in values exists, then no problem exists. Second, ethical problems can be found both in the subject matter of the research problem and in the manner in which the research will be conducted. A current example would be the use of aborted fetal tissue for research; both the abortion and the use of the material are subject to conflicting values. Further in this case, sensitivity to the conflict in values will not lead to a solution, as the groups having values at both extremes will not work to resolve the issue. The ethical conflict can involve both personal and professional components. Individuals can experience conflict between these two elements of life. As the complexity of the research increases, so too can the questions relating to ethical behavior. Often a single research problem can raise multiple questions about proper conduct necessary to protect human rights, as well as protect the integrity of the data. Finally, many ethical decisions are not yes or no, black or white; rather, judgments about proper ethical conduct often lie somewhere in between. Yet, many individuals expect that a clear, simple, dichotomous yes or no decision can be made in all research situations. This has led to the need for increased integrity on the part of researchers.

Ethics and Integrity

At the end of World War II, the discovery of abuse by Nazi scientists led to the development of the first ethical codes of conduct applied to researchers. The ethical issues in science and research have raised increased concern as researchers try to ensure that their studies are directed toward worthwhile goals and that the welfare of their subjects and research associates is protected. Such concern is evidenced by the number of codes that have been developed, beginning with the Nuremberg Code of 1947. Subsequent internationally accepted codes include the Draft Code of Ethics, World Medical Association, Geneva, 1961 and the World Medical Association Declaration of Helsinki (1964; rev. 1975). Congress has passed several regulations that specify criteria for scientific research and require researchers to report any incidence of misconduct, for example, 42 C.F.R. Chapter 1, Part 50; 45 C.F.R. Subtitle A, Part 46; and 9 C.F.R. Chapter 1. Copies of sections of these codes and regulations are found in the appendixes. See appendix A for the Nuremberg Code, appendix B for the Helsinki Declaration, appendix C for 42 C.F.R., appendix D for 45 C.F.R., and appendix E for 9 C.F.R.

Kimmel (1988) discussed ethical theories and standards used as research guidelines. He observed that since scientists and others can differ in the general approaches used for ethical decision making, it is important for individuals to be aware of the approach they use when faced with a moral problem. The issue of ethics is further complicated by a disagreement among philosophers as to the precise scope of ethics. Some philosophers accept the notion of normative ethics, or a set of principles that guide human behavior. Others argue for a metaethical approach, which focuses on the analysis or logic of moral concepts rather than on a systematic theory of ethics for daily living. As described earlier, normative ethics sets forth standards used to judge the morality of actions, whereas metaethics considers the process that people use when they make moral decisions.

In spite of the philosophical debate as to whether a normative ethical or metaethical approach should be adopted, the government and professional organizations have rushed to embrace a normative approach. The result of governmental activity is that series of rules, regulations, and procedures have been put into place to guide research, especially scientific investigations involving human or animal subjects. The intent of the government and professional organizations is to provide rules that apply to all research.

According to this approach, judgments, if challenged, may be justified ethically by showing they conform to one or more rules or norms. A rule is a general statement that actions of a certain type ought (or ought not) to be taken because they are right (or wrong). At an even more fundamental level, ethical principles serve as foundations for the rules and norms. Finally, there are ethical theories that consist of systematically related bodies of principles. At the heart of this system lies the concept of a basic ethical principle, which justifies the many particular prescriptions for and evaluations of human actions (National Commission 1976, 1977). To develop a system of ethics, one begins with a theoretical basis. Principles, rules, and norms for behavior are developed by using this ethical theory. The behavioral norms represent the most specific aspects of an ethical system. See figure 1.1 for a representation of an ethical system.

Proponents of this ethical system believe that a fundamental ethical principle can be formulated. This is a principle that exists within a system of ethics and is taken as an ultimate foundation for any second-order principles and norms. A fundamental principle represents a "pure" or "basic" ethical value because it is not derived from any other statement of ethical values. Thus, an ethical norm is a statement manifesting that actions of a certain type ought (or ought not) to be taken. An ethical norm statement contains the words *should*, *ought*, *must* or *forbidden*. Examples of ethical norm statements include the following:

> The experiment should be so designed and based on the results of animal experimentation and a knowledge of the natural history of the disease or other problem under study that the anticipated results will justify the performance of the experiment. (Nuremberg 3)

> The experiment should be conducted only by scientifically qualified persons. The highest degree of skill and care should be required through all stages of the experiment of those who conduct or engage in the experiment. (Nuremberg 8)

> Biomedical research involving human subjects must conform to generally accepted scientific principles and should be based on adequately performed laboratory and animal experimentation and on a thorough knowledge of the scientific literature. (Helsinki I. 1)

> Biomedical research involving human subjects should be conducted only by scientifically qualified persons and under the supervision of a clinically competent medical person. The responsibility for the human subject must always rest with a medically qualified person. (Helsinki I. 3)

> Respect for persons incorporates at least two basic ethical convictions: First, that individuals should be treated as autonomous agents, and second, that persons with diminished autonomy and thus in need of protection are entitled to such protection. An autonomous person is . . . an individual capable of deliberation about personal goals and of acting under the direction of such deliberation. (National Commission 1976)

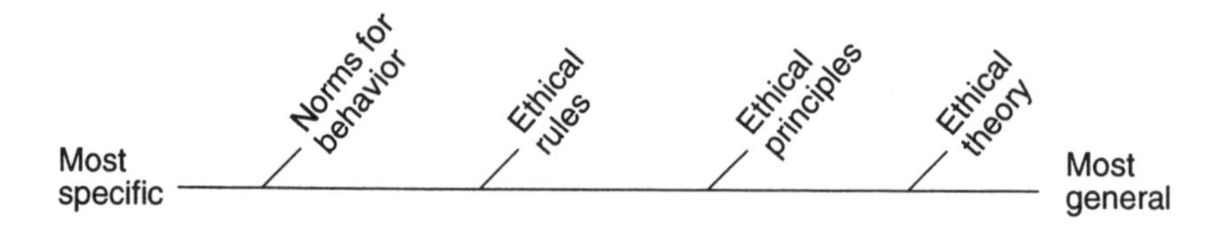

Figure 1.1 Schematic representation of a system of ethics.

The question remains, however, can one fundamental ethical principle serve in all situations? The current need for Acquired Immunodeficiency Syndrome (AIDS) research and policy serves as an example of the issues raised when dealing with ethics in research and science. The ethical norm statements above indicate that research must be based on a sufficient background of animal research and adequately performed laboratory research before human subjects are introduced. However, the desperate need for medical treatments arising from the pressure to do *something* caused the governmental agencies to issue guidelines allowing drug companies and physicians to bypass the traditional procedures used when testing medicines for safety and efficacy.

Some individuals argue that safety is not a major issue when people afflicted with a fatal disease are involved, and they should be provided with any and all opportunities, however risky, to receive whatever treatments they wish to endure. Even in such dire circumstances, however, highly toxic drugs can potentially reduce their life span and/or affect their quality of life.

An equally important question arises. Absent the use of traditional methods of scientific evaluation, how will the quality of these clinical trials be determined? Should the traditional ethical norm statements be discarded in the face of pressures to do something to stop the advance of a fatal communicable disease? These same ethical issues were present in the breast cancer study fraud highlighted earlier. Research is used to influence and guide policy development, so the quality of research is of paramount importance to decision makers. Our responses to the AIDS epidemic have been reactive, which in turn compromises the government's ability to develop effective policy. Thus, the abandonment of traditional scientific methods has a great impact on society as a whole, and increased risk from poor research is not limited to only those individuals with AIDS who are electing to receive questionable treatments (Kaemingk and Sechrest 1990). The proper scientific foundations for procedures being developed for future generations of individuals needing treatment must also be considered as a part of the ethical consideration when traditional methods are disregarded.

Returning to the issue of the war crimes trials in Nuremberg, a great deal of attention was focused on the unethical scientific experiments conducted by Nazi scientists. For example, Nazi scientists recorded how long it took people to die when they were immersed in ice water. No amount of punishment, ethical norm statements, or other actions will bring back to life those individuals who suffered through the horrible experiments. But the question remains, what should be done with the data? Should all the data from those studies be discarded because of inhumane research, or should data that provides a legitimate contribution to knowledge, if any exists, be kept and used so those subjects did not die in vain?

During the 1930s and 1940s physicists worked on basic research that led to the development of nuclear fusion and fission. They discovered atomic energy, which led to the development of the atomic bomb and even more powerful nuclear weapons. Pure researchers would argue that the development of nuclear energy is amoral (i.e., neither good nor bad). The resulting atomic bomb and its use on two Japanese cities during the war has been described as both good and bad. Some argue that such a horrible weapon should never have been used; others argue that without the use of the atomic weapons, the war would have lasted much longer and thousands more would have died, far exceeding the number killed by atomic bombs. In fact, more deaths and destruction occurred from the incendiary bombs used on Japan than from the two atomic bombs, yet the moral aspects of this bombing have not been questioned. Nuclear energy has since been used to provide electrical power in peaceful settings and has been hailed by some as an environmentally friendly source of energy.

The discovery of any new knowledge raises ethical questions that must be answered. In the case of new discoveries such as nuclear energy or genetic engineering, is the discovery of new knowledge that has potentially deadly results morally neutral? Must the potential of new knowledge for deadly results be weighed against its potential for beneficial results? Finally, do the "pure" researchers who identify such new information have a moral responsibility concerning the release and use of their findings?

Answers to these and similar questions related to science and research depend on your philosophical approach to ethics. If a code of ethics is to be accepted, what will provide the standards by which the correctness and incorrectness of conduct can be judged? Who will determine the standards to be used when judging ethical decisions? The prevailing rules of a culture have been rejected from time to time as appropriate standards of right and wrong behavior. Keeping this problem in mind, is there a fundamental ethical or moral principle that can guide all decision making? Can a series of ethical rules apply to all situations faced in science and research? In life? Should all scientists be required to take a course in ethics? Should ethical decision making be a part of all professional curricula? What is the role of ethics in research and science?

Summary

Ethics is one of several disciplines that study human behavior. Ethics views behavior from the perspective of right or wrong, good or bad. The system of ethics used by professional organizations and governmental agencies is that of rule utilitarianism. Its goal is to develop a system of ethical statements that has universal application and that delineates which behaviors are right and which are wrong. This approach evolved from publicity concerning reports of fraud in science. Studies have indicated that problems exist in both laboratory research and clinical research. Some problems were caused by researcher dishonesty. Fewer problems resulted from poor organization and improper techniques. Sensitivity to potential ethical issues in research is required, but increased sensitivity alone will not eliminate the problem. Dishonest individuals must be disciplined and ousted from the research professions, and opportunities for fraudulent research practices must be minimized as much as possible.

CHAPTER 2

BEGINNING THE RESEARCH PROJECT

Any individual who develops a research idea, whether it be for a class assignment, personal research, or a grant submitted to a funding organization, will follow a series of specific steps. First these steps will constitute the proposal, then they will be used to draft the research project. The investigator begins by formulating a problem that will be studied and ends by publishing the results, whether it be in a class paper, journal manuscript, conference presentation, or report to a granting agency. The researcher must make one or more ethical decisions at each step of the research process. This chapter presents types of ethical decisions that appear during the development of a research project, including the formulation of a statement of the problem and the role of Institutional Review Boards (IRB).

Developing the Problem

Ethical situations in research can be observed as early as when the researcher thinks about a research problem. The ethical decisions become essential as researchers develop the statement of the problem that will be studied. Ethical problems can occur with the decision to conduct research as well as with the decision not to conduct research. Researchers might decide not to pursue a research topic for several reasons: the study might invade someone's privacy, too much time will be required to get subjects and clearance from an IRB, or the necessary procedure might be too controversial, time-consuming, or costly. In each case, researchers might be making life easier for themselves, but they may neglect to conduct research that will be in the best interest of society or the research participants. At this point they may ask themselves, Why conduct research anyway if you have to put up with all the problems associated with developing a topic, getting IRB approval, and dealing with subjects while collecting the data? Is the hassle worth the trouble it causes for me? For society?

Pring (1984) believes there is a need for knowledge and that, without access to knowledge, society would be worse off: ''Maybe the right to know . . . will lead to undesirable results. Maybe it will clash with other principles. . . . The main thing however is that, as in most moral matters, whether the right should be given and over what area it should be extended need to be argued in the context of particular cases'' (8-18). Pring argues that a need exists to determine specific cases in which knowledge is made available and that moral decisions should be made regarding these cases. The appropriate time for this type of discussion is during the project's formulation. If researchers fail to do so and decide that a research project is not worth the effort, they have merely traded one problem for another, unaware that they have made an ethical decision. Thus, ethical decisions are made before the problem is framed for study, and decisions continue to be made even after the investigation is completed. The following case study illustrates one type of ethical problem investigators may face when framing their research.

CASE 2.1

HONESTY WITH CORPORATE SPONSORS

A university professor plans to submit a research proposal that will receive considerable support by a corporate sponsor. This research could lead to the development of new manufacturing processes that would benefit the sponsor. The grant for this research will not include contractual agreements whereby the professor becomes an employee of the corporation. Basic intellectual and scientific ethics dictate that the data and experimental method used should be open and published. However, in the corporate world, trade secrets, business ethics, and other considerations require that some information be kept confidential. The findings of this proposed research will likely result in information that will give the corporate sponsor an advantage over its competitors and would fall into the category of trade secrets. In addition, the researcher feels that publication of this research will likely enhance her reputation as an authority in her field, and she wants to publish the results of this study.

How should the researcher deal with this situation? Should the researcher wait to see if any stipulations regarding publication accompany the grant? Who owns the data? Who owns the findings? If no statements about publication are present, should she ignore the problem during the formation of the proposal and publish her findings without considering the wishes of the corporate sponsor once the study has been completed? What are her ethical duties to her profession? Her university? The corporation? To future scientists who try to secure grants from the corporation? Does this situation fall within the earlier proposal made by Pring?

Selection or Rejection of a Topic

Once an individual chooses to engage in a research project, ethical decisions present themselves regarding the choice of topics studied. For example, since many journals will publish only articles that report statistically significant results, investigators often feel extreme pressure to develop a study that will produce statistically significant results. This pressure might be present for master's or doctoral research if the institution requires that an acceptable thesis or dissertation contain statistically significant results. It is also present for graduate students, faculty members, and others who are under pressure to show a record of publication and who desire to use this work to meet their publication requirements. Likewise, many professional meetings will not accept presentations that lack statistically significant results. Researchers who depend on grants to fund their research know that failure to produce statistically significant findings can result in the loss of future funding. Consequently, an investigator is often pressured to choose problems that are almost certain to produce such findings. The topic selected can be one known to include questions that produce statistical significance, or the researcher can choose to investigate multiple questions using large numbers of subjects that will, by chance alone, likely produce significant results.

Statistical significance differs from practical significance in that it means the results will occur from factors other than chance. On the other hand, practical significance looks at the utility of the results and may include such factors as cost, time, risk to others, and availability in its determination. Statistical significance may not result in practical significance and, in fact, the lack of statistical significance may have practical significance. If research findings show no difference in the efficacy of two drugs used to treat seizures or mental illness, but there is a difference in retail prices, then there is an issue of practical significance for the consumer. Does this raise ethical considerations for the parties involved with the research? Does your view of the ethical problem change if one product is more profitable than the other?

What, if any, ethical issues arise when a researcher avoids legitimate problems because the investigation might not produce statistically significant results? Does the policy of a journal limiting its acceptance of manuscripts to only those having statistical significance produce an ethical problem? Does a profession have a moral obligation to provide information that two drugs, educational programs, or other factors are equivalent?

Multiple or Single Ethical Issues

With ready access to considerable computing power, the nature of many investigations has changed. Evaluating large amounts of data and multiple hypotheses in a short time is now possible. Consequently, many researchers no longer develop and conduct single-issue studies. When conducting a complex study, an investigator is unlikely to be successful in isolating the ethical question in the study, as multiple issues usually arise. Even in single-issue studies, multiple ethical issues are present if the proposed study is carefully evaluated by the investigator. Ethical questions arise as the proposal and statement of the problem are evolved. How are subjects selected or omitted? Must subjects be labeled and, if so, how? What experimental procedures will be used? What impact will the procedures have on the subjects, confidentiality, collection and storage of data, release of findings? And how will the authorship of manuscripts be resolved? Such issues will be explored in greater depth later.

Deciding what to include in the research problem can be another ethical issue. Should the investigator include topics that fall outside his or her area of expertise? Must the investigator be neutral about all topics included in the study? Consider the following case study.

Case 2.2

The Use of Added "Variables" in Research

Researcher Doe is a consultant for the MetaValue Athletic Supplement Company, Inc. He receives both a retainer fee and a royalty based on sales of the supplements he helps develop. He knows that sales will improve if he can prove that MetaValue Supplements will improve performance. So Doe looks for opportunities to include MetaValue products in every study that he conducts to evaluate athletic performance. Doe decides that it is quite simple to add one more variable—namely, MetaValue—and collect all the data possible. He adds MetaValue to all his studies, even when a project is funded by government or agency grants. He does not tell the funding agencies that he is adding the supplements, nor does he mention MetaValue in his analyses. MetaValue data is analyzed separately and reported only in advertisements or professional journals when outstanding results occur. Who is injured by this practice, Doe reasons, since the MetaValue data won't be included in the research report sent to the grantor? After all, they got what they paid for.

Recall, one purpose of the scientific method is to remove any investigator bias. Is Doe violating the principles of the scientific method? Does his choice affect ethics and integrity in research? Does it represent a legal breach of contract? Is a conflict of interest present? Does Doe's choice represent a violation of professional ethics?

Selection of Subjects

Although it sounds like a simple issue, the selection of subjects can raise several ethical questions. Several subpopulations have not been used extensively because subjects and useful data are not as extensive for these groups as they are for others. It's been said that the most widely used subjects are male army recruits, male college freshmen, and white rats. Groups that have not received a great deal of study and have a dearth of data are females, older adults, children, and individuals with disabilities. Years ago, young adult males were more readily available for use as subjects because they made up most of the college student body and military personnel. However, although demographics have changed as female college enrollments and military enlistments have increased greatly, young adult males remain the subjects of choice in many instances.

Perhaps the major reason for this reliance on young adult male subjects is that often it is easier to use them than members of other subpopulations. When using young adult females as subjects, the issue of pregnancy may appear. The researcher may be asked by an IRB to add provisions, such as screening tests, that will eliminate any subject from the study who is or might be pregnant. Likewise, IRBs may require that all subjects over the age of 40, regardless of their health status, knowledge of the risks involved, and readiness to participate, have a complete physical examination before the study begins. Such requirements will add considerable cost and complexity to the study, including the issue of who pays for the examination. Often investigators, when faced with these requirements, choose to limit their subjects to young adult males so they will not be faced with these additional costs and procedures. Does the elimination of females, older adults, and children as subjects pose an ethical problem? If so, for whom does the problem exist: the researcher, the IRB, or the funding agency?

Once the subjects have been selected, the question arises as to how you designate the groups and whether you tell the subjects which group they are placed in. Labeling individuals can create serious experimental problems. Rosenthal and Jacobson's 1968 book, *Pygmalion in the Classroom*, received immediate and widespread attention in the education community. It was the result of a 10-year research project funded by the Division of Social Sciences of the National Science Foundation, which investigated the effects of interpersonal, self-fulfilling prophecies. Their results showed that one person's expectation for another's behavior could serve as a self-fulfilling prophecy; children classified as bright, though they were of average ability, were regarded as bright by their teachers and their test scores improved correspondingly. The actual difference between these "bright" children and normal children existed purely in the minds of the teachers.

Considering such findings, the use of labels for subjects has ethical concerns. In social intervention research, it's a common strategy to achieve early identification of individuals at risk and modify their environment in attempts to reduce the risk. Often, this procedure means that others such as teachers, therapists, parents, and perhaps peers must be involved. If such individuals need special services, then they must be labeled before the services will be provided (e.g., mentally retarded, mentally ill, or developmentally delayed). The use of such a label, however, increases the possibility that interpersonal self-fulfilling prophecies will result, producing a negative effect on the at-risk subject. Thus, an ethical problem exists in the choice between helping the at-risk person by assigning a negative label and not providing the person with the assistance necessary by failing to assign a label.

Several solutions have been proposed in attempts to avoid labeling problems; however, these solutions also raise practical and ethical problems. One suggestion, for example, is to use the schoolroom or regional ''broadcast'' design. Using this intervention approach, groups of people, such as schoolrooms, regions, or institutions, receive the treatment, rather than only the targeted at-risk individuals. This raises problems of logistics and obtaining parental consent from parents of children who are not at risk. Finally, to determine the impact of the intervention on the target groups, subgroups must still be formed to allow for follow-up evaluations. These procedures serve to increase the cost of the project and, in the end, individuals must still be identified as at-risk and labeled. Is it ethical to provide the service for individuals both at risk and not at risk, spending large amounts on those who will likely not benefit? Is it more ethical to limit the services to those individuals at risk, even if labeling is required? Does labeling fail to raise an ethical problem, since the subjects have likely already been labeled by those observing and interacting with them?

Labeling is but one way in which participation as research subjects can cause difficulties. Intervention studies that seek to change behaviors and produce positive changes in individuals can also cause difficulties, not only for the person whose behavior is changing but also for those who have close relationships with that person. The use of intervention research raises ethical questions that must be answered. Consider the following example, for instance.

CASE 2.3

RESEARCH THAT CHANGES RELATIONSHIPS

A researcher is planning a study regarding health promotion and wellness that will investigate the roles that family members adopt toward each other that may correlate with healthy or unhealthy lifestyles. Developing an intervention study, the researcher decides to use a technique that will change one or more of the family members' attitudes toward their role in the family in the hope that this will facilitate the adoption of a healthier lifestyle. The researcher is well aware of data indicating that changes in roles by family members can cause strife within the family. In some cases, the use of intervention techniques can damage family and interpersonal relationships.

Should the researcher individually make the decision whether to use this technique? Does the potential for benefit to an individual outweigh the potential for stress and harm? Who should decide? Knowing this type of research may cause turmoil within the family, what is the researcher's responsibility to the subject and to other family members? When reporting the results of such a study, is it ethical to report only the positive changes in the individual without telling of the problems the change in behavior caused others?

Institutional Review Boards

Early codification of ethics for use by professionals engaged in scientific research, as seen in the Nuremberg Code and Helsinki Declaration, emphasized the role of the scientist in determining the safety of the subjects. Scientists were required to exercise "the good faith, superior skill and careful judgement" (Nuremberg 10) and to terminate the experiment if injury, disability, or death of the subject was likely. The Helsinki Declaration emphasized that "the importance of the objective is in proportion to the inherent risk to the subject" (Helsinki I. 4) and that research projects "should be preceded by careful assessment of predictable risks in comparison with foreseeable benefits to the subjects or to others" (Helsinki I. 5). Both codes placed the responsibility for making such determinations on the investigator. Some of these principles have changed following proposals made by government study groups. During the 1970s, the National Commission for Protection of Human Subjects of Biomedical and Behavioral Research (National Commission 1976, 1977, 1978) published a series of reports with the result that much of their work was incorporated into federal regulations. The 1978 report expressed their concern that Institutional Review Boards (IRB) should be made a part of the research granting process, and it listed recommendations concerning their establishment and function.

Today the basic IRB policies and requirements are set forth in 45 C.F.R. Subtitle A, §46.101 et seq. Present government regulations limit the required review activities of the IRB to research funded by selected government agencies such as the Department of Health and Human Services (DHHS) or National Institutes of Health (NIH). This is a change from earlier regulations that required review of all research proposals developed at institutions who received any government support for research. However, many universities and other agencies continue to extend their requirement for IRB reviews to research proposals in situations that far exceed the scope required by government regulations. Many institutions give their IRB the power to review all research involving human or animal subjects, thus enabling the IRB to approve or disapprove all such projects conducted by researchers related to the institution. An IRB review may require modifications in the proposed research procedures that will change the thrust of the proposed research. As described earlier in the discussion about the selection of subjects, costly and stringent requirements are often imposed on a research project that practically eliminate certain populations from use as subjects. These requirements frequently impact populations, such as the elderly, disabled, and females, where information required to make important policy decisions is lacking. With elderly subjects, the additional factors that usually preclude their use as subjects relate to the requirement that all individuals must have expensive physical examinations before they can serve as subjects. With females, the requirement that causes investigators to discard them as subjects usually focuses on the potential for pregnancy, even though pregnant women have been, and continue to be, participants in stressful international athletic contests with no evidence of harm accruing to either the women or their unborn infants. Disabled subjects are often eliminated because of the extra needs associated with such individuals, the fact that they are unable to appreciate the risks they may experience as subjects, the added costs associated with their inclusion, or the feeling that they are medically high-risk individuals who should not participate as subjects in research projects.

Another ethical issue regarding the use of IRBs is the membership requirements associated with the boards. According to DHHS regulations (45 C.F.R. Subtitle A, §46.107), an IRB must have at least five members with varying backgrounds, and they cannot be all men, women, or members of one profession. Each board must contain one member in a nonscientific area (e.g., lawyer, ethicist, member of the clergy) and one member who is not affiliated with the institution and not a relative of a person affiliated with the institution. IRBs cannot have members who are participating in a project under review or who have a conflict of interest. Because of special membership requirements, IRBs often may invite individuals with competence in special areas to assist in the review, but they will not allow these individuals to vote on the project under consideration. There is no information about how often IRBs invite such experts or how their ideas are used by the board.

A series of practical and ethical questions arise about the nature of the IRB review and approval processes. Examples of these issues are as follows: What primary ethical concern does the IRB represent? The welfare of the institution? Protection of subjects? Implementing government regulations? An IRB review should balance the suitability and importance of the research topic under review with the risks involved for the subjects. Consequently, the IRB must pay attention to the design and methodology of the study. Considering this task, does the membership requirement stated in federal code adequately constitute an IRB that can review a sophisticated topic that only a few investigators in the field could understand? Should the IRB review be limited to safety of the human and animal subjects? Should the IRB have a goal of eliminating "bad" research? If so, who or what determines what constitutes "bad" research? Should the IRB review all proposals that use human and animal subjects in research activities? Should the IRB be in the business of revising research projects when IRB members either don't understand or don't agree with the procedures proposed in situations where reasonable scientists may differ? Should the IRB attempt to eliminate all risk or only unreasonable risk of harm to subjects? If only unreasonable risk of harm is eliminated, how is reasonable risk determined? How much weight should an invited expert's participation have in decision making? Should the IRB impose such extensive and expensive requirements that effectively eliminate certain populations as subjects because of the difficulty a researcher will experience in meeting the requirements? Should IRB reviews be limited to only a consideration of funded research?

IRB reviews require the vita and presence of the investigator, or the investigator's representative (e.g., dissertation advisor), when reviewing a proposal. IRBs have reportedly disapproved studies that would otherwise have been acceptable based on their evaluation of the investigator's qualifications. Are IRBs equal to the task of evaluating an investigator's qualifications to conduct research? If so, what should they consider in their evaluation? Reviews of manuscripts submitted for publication in refereed professional journals are subject to "blind" reviews in which the identity of the author or authors is not known to the reviewer. This procedure is used in an attempt to eliminate any bias or "halo effects" through which well-known authors have their work accepted by the weight of their reputations while unknown authors may have their work rejected. The principle inherent in this practice is that manuscripts should be reviewed on their scientific merit without regard to the authors' reputations. Should IRB reviews be conducted using the blind review procedure so that the review is limited to the scientific merit of the proposed study?

In summary, several concerns about IRB reviews of proposed research have been raised. Kroll (1993, 42-43) writes:

> In another example of IRB jurisdiction in non-scientific aspects of a study, the board requires an estimate of the amount of time each human subject will be expected to devote to the study. One can only surmise that the IRB is empowered to decide what a reasonable time commitment is for human subjects and to disapprove those studies it determines take too much time. IRBs also require information about the amount of any compensation provided to subjects. Such a provision must lead to decisions about appropriate amounts of compensation, although the original intent was probably to discourage subjects from accepting more than minimal risks in order to earn financial stipends. Numerous related examples could be cited, but it seems safe to suggest that IRBs as well as policies of federal agencies often intrude on researchers' autonomy and make value judgments about the conduct of science, which can be challenged as being beyond their scientific expertise or intended charge.
>
> And it is precisely to this issue that most complaints about institutional review boards have been directed. The IRB often makes recommendations and conditions for approval that transcend simple attention to possible risks and safeguards. The IRB often assesses the scientific worth and methodology of a study under review. So too sensitivity to the needs of society extends to *prescriptions* for research methodology. Consider the priority announcements for research involving biomedical and behavioral

> studies of diseases and disorders. The NIH/ADAMHA policy is that applicants must include minorities and women in study populations "so that research findings can be of benefit to all persons at risk." If women or minorities are not included or inadequately represented, "a clear and compelling rationale should be provided." If such information is not clearly provided, the initial review group is to request such information; if it is not provided, the application is deferred or returned to the applicant.

While the decision to review and reduce unnecessary risk to human subjects is ethically sound, many IRB functions go beyond this original charge. Although the NIH/ADAMHA policy requires the inclusion of women and other groups in studies, IRB-imposed requirements often serve to eliminate the possibility that these populations will serve as subjects in some research projects. Research today has become political at the funding level and at the institutional level. Ethical questions about the effect of political approaches upon scientific procedures are many, and individual researchers may not have much choice in the matter. Does the introduction of political factors into the review process create ethical problems? If so, for whom? The researcher? The institution? The IRB members? Should human subject review procedures be revised to eliminate any political considerations? If so, what revisions should be made?

Summary

Ethical problems can appear as soon as a researcher begins thinking about a problem, defining it, and developing the methods that will be employed to study the problem. Even the decision not to pursue a problem involves an ethical decision by the scientist. The selection and treatment of subjects raise several ethical questions. Not the least among these is the population from which they are drawn. Should subjects be limited to young adult males, or should the subject pool include other groups? When research uses human subjects, labeling of subjects becomes an issue for the researcher. Current government regulations demand review of human and animal studies proposed for funding by DHHS and NIH by an IRB associated with the institution where the research will be conducted. Many institutions have expanded the role of their IRBs to review all studies involving human and animal subjects. The composition, purpose, and involvement of an IRB also raise ethical questions in research.

CHAPTER 3

DEVELOPING YOUR METHODS

In the previous chapters, we discussed potential ethical problems associated with the decision whether or not to conduct a research investigation and, if so, which subject populations should be used. Once you have decided to pursue a research topic and have determined the characteristics of the subjects you will use in the project, you can begin to develop the research methods and analyze the data you obtain. The research methods component of a project includes securing your subjects, developing procedures, choosing the tests and protocol that will provide the data, and deciding how the data will be analyzed. Each of these components requires an ethical decision, so the researcher must be aware of the ethical issues that arise during this phase of the research. Typical questions associated with the methods used in research are many and varied. Decisions made during the design of a research project are critical to the conduct of a responsible study according to the scientific method and to the protection of subject rights and welfare. Without the use of proper methods, the study will not meet accepted scientific criteria and its utility will be limited.

Securing Subjects

As mentioned in chapter 2, the most frequently used subjects have been white rats, army recruits, and college freshmen, although not necessarily in that order. The reason for using white rats in biomedical research is obvious; possibly the same thing could be said for the use of army recruits as subjects. However, why are college freshmen so widely used as experimental subjects? Typically college freshmen are enrolled in large lecture sections of introductory classes, and they are required by their professors to participate as subjects in an experiment before they can receive a grade in the class. Sometimes students are given extra credit or bonus points for participating in experiments. The reason usually set forth to justify this requirement is that such participation will introduce the students to scientific research and instill an appreciation for research. If that's an instructional objective, why not assign groups of students to conduct a study of their own using the entire scientific method during the project rather than coercing their participation as subjects for research projects that benefit the professor?

Most research codes of ethics require that subjects give their informed consent before serving as subjects and, as a part of the consent, the subjects must be assured that their nonparticipation will not impact on them in any negative manner (see informed consent discussion that follows). Under the most current regulations, if students participate in activities that are a regular, ongoing course requirement, no IRB review is necessary, and students are not required to sign informed consent documents. Several ethical dilemmas are raised by this state of affairs. Does a course requirement that compels students to serve as subjects to obtain a passing grade for the course present an ethical question? Should such requirements face IRB review? Should these student subjects be treated differently by the researcher than other subjects if the results are used for more than fulfillment of a course requirement? Should such course requirements be subject to IRB review and include the same informed consent standards as other research? If the researcher uses subjects who are coerced, does that taint the data obtained from the study? Should the researcher acknowledge that coerced subjects are used in the study in any publications that may result?

Consider further some potential ethical problems present in the situation in Case 3.1 where a professor requires students to participate in research studies that will benefit the professor as a condition for receiving a course grade. Does this practice represent a conflict of interest for the professor? Should a professor be able to require students to participate as subjects in research projects developed to benefit the professor, or should these studies be standardized experiences that supplement the class from which the professor derives no benefit? The nature of benefits gained by the professor could take several forms. For example, it might be that the professor is receiving grant money for funded projects and the students are used for subjects in these studies. According to the policy of many universities, professors may receive increased salary through promotion and merit increases flowing from publication of the research conducted in courses where the students are required to participate. Or the professor may be developing a new commercial product and may use the student research requirement to provide subjects for product development research.

CASE 3.1

CAPTIVE SUBJECTS

Dr. Fiz Whiz is an exercise physiologist who has just signed a contract with the makers of ErgoStat, a food supplement touted as a source of quick energy and a performance enhancer. According to Dr. Whiz's contract, he must provide data that the manufacturers can use in advertisements that will increase ErgoStat sales. Subsequently, Dr. Whiz redesigns his exercise physiology classes to include a requirement that all students must participate as subjects in a research project before they can receive passing grades in the course. He designs these research projects and assigns students. All research projects requiring student participation relate to the effect of performance enhancement products and include ErgoStat as one of the variables studied. Dr. Whiz sends any data showing ErgoStat in a favorable light to the company for use in their sales campaign. Furthermore, Dr. Whiz submits manuscripts with data from these class-required projects that meet publication criteria to national journals for publication under his name.

If the students' required participation as subjects reflects an educational experience germane to the course, should the professor be allowed to publish the results of these "educational" research projects? If the research will better the professor's status through funding, promotions, raises, or product development, should the students be paid for their involuntary participation as subjects in the research projects? Should potential conflicts of interest in this situation be ignored?

When using subjects other than readily available persons such as college students and army recruits, volunteers are the next most frequently used subject pool. No matter how the subject pool has been selected, the problem of assigning subjects to the control and experimental groups must be faced. Assignment of subjects must be made in such a manner that research and statistical biases are eliminated or minimized. Random assignment of subjects appears to be the fairest and most unbiased method for determining who will participate in a research project and what treatment each individual subject will receive. Usually random assignment is made using either a lottery or a table of random numbers to make the assignment. Even though the use of random selection and assignment of subjects appears to be a fair approach, ethical issues arise.

For example, should all subjects who participate in a research study receive some direct or indirect benefit from their participation? In a medical study where a medication that will likely promote better healing is used, a benefit to the experimental group can be projected. Likewise, if the control group will receive the standard treatment, these subjects will also accrue a benefit. However, consider the experimental situation in which the control group will receive no treatment while the experimental group receives a treatment; the following questions arise from this situation. Are the subjects in a control group unfairly denied benefits that might arise from participation in the experimental treatment? Conversely, will the experimental group be unfairly subjected to unproven, questionable, and potentially harmful procedures?

Related, and perhaps of greater concern, are those studies in which subjects undergo potentially detrimental treatments or situations. Consider, for example, studies requiring long periods of immobilization to produce muscular atrophy or other types of deconditioning. These studies are used for research to develop better therapeutic treatment to recondition patients whose limbs were immobilized by injury or disease or to study the potential effects of prolonged flights into space on astronauts. Can scientific studies that expose subjects to potentially damaging procedures or potentially harmful situations be justified ethically, even if subjects volunteer? Often subjects participate in such research without receiving any apparent direct benefit. Should a requirement be imposed on investigators that all subjects must receive a benefit from their participation before the research project will be approved? If so, what is the nature of the benefit that must be received? Will payment for participation in a research project or the fulfillment of an altruistic feeling that accompanies helping others meet such a requirement? If payment is acceptable, how much money should a subject receive for participation? Should the payment be commensurate with projected or actual risk? How can the presence of an altruistic benefit be determined or proven? Is the requirement that a research project must produce a benefit to others, without a requirement of direct benefit to the subjects, sufficient to meet ethical standards required for approval? Must all research meet a specific requirement that a direct benefit resulting from the study in question be demonstrated before the proposal will be approved? If so, this requirement may eliminate research that is conducted for its own sake. ''Pure'' research conducted to increase general knowledge may be prohibited.

If such a requirement is imposed that leads to a reduction in ''pure'' research, a different set of ethical questions is raised. What is the ethical justification for prohibiting research that increases general knowledge, thus reducing the general pool of knowledge that may be used to produce future benefits for society? Is a requirement that all research must have immediate and practical utility ethically defensible?

Informed Consent

The concept of informed consent comes to research from tort law defenses used in the legal system. Lawyers searched for ways to transfer the risk of loss from their clients to others. They developed informed consent to show that the injured parties had agreed to assume the risk of injury that was present in their participation and that their client therefore was held harmless. In the legal system, the assumption of the risk doctrine requires that two conditions be present before the injured party will have assumed the risk of injury and thus leave the other party harmless: (1) the injured party must have knowledge of the risk and all potential harm that may come from participation in the activity and (2) the injured party has voluntarily agreed to participate in the activity after having such knowledge.

The basic elements of informed consent regarding risk in research studies are set forth in 45 C.F.R. Subtitle A, §46.116, General Requirements for Informed Consent, and include the following:

- A description of any reasonably foreseeable risks or discomforts to the subject (§46.116(a)(2))
- For research involving more than minimal risk, an explanation as to whether any compensation and an explanation as to whether any medical treatments are available if injury occurs and, if so, what they consist of, or where further information may be obtained (§46.116(a)(6))
- A statement that participation is voluntary, refusal to participate will involve no penalty or loss of benefits to which the subject is otherwise entitled, and the subject may discontinue participation at any time without penalty or loss of benefits to which the subject is otherwise entitled (§46.116(a)(8))

The National Research Act, Public Law 93-348 (1974), presented a more expansive set of elements that should be included in the informed consent document. In addition to those requirements listed in 45 C.F.R. above, the following statements must be provided to the subjects:

- A description of any benefits to the subject or to others that may reasonably be expected from the research
- A disclosure of appropriate alternative procedures or courses of treatment, if any, that may be advantageous to the subject
- A statement describing the extent, if any, to which confidentiality of records identifying the subject will be maintained

Inasmuch as many of these requirements come from tort law where they are used to shift the risk of injury or loss from one party to another, what is their role in research? A further series of questions must be asked. Is this just another risk management tool? If so, which party is protected by the informed consent requirement? The researcher? The institution? The subject? In the legal context, informed consent usually applies to physical harm because mental injury is extremely difficult to prove in tort law. In the research setting, is the IRB's inquiry of harm limited to physical injury, or does it include potential psychological injury such as loss of self-esteem or self-confidence? If so, should a distinction be made between potential physical and psychological harm? Should psychological injury be defined? How does a researcher show whether mental anguish is or is not likely to be present during the research?

Many research studies involve children or individuals otherwise lacking the capacity to be classified as adults (e.g., mentally retarded). Such individuals, called minors, do not have the legal capacity to understand or appreciate any risks that may be present through their participation as subjects in a research project. In these situations, other individuals must give consent before the individual without adult capacity can participate in the research project. What legal and ethical ramifications flow from this situation? In tort actions, courts have held that adults cannot release their child's legal rights when the adult signs the release and waiver form often required for the child's participation in sports camps, on athletic teams, and in other situations requiring such documentation. Courts hold that the adults may release their (the adults') right to sue, but the child or the child's estate retains the right for legal relief. Therefore, if a parent or guardian would sign an informed consent statement, what is the legal status of the document if injury should occur during the minor's participation in a research project? If a parent or guardian is permitted to sign an informed consent document for a minor, should the issues that a parent or guardian must consider before giving consent for their child's participation as a subject be standardized or left to the parent or guardian's own discretion? Should the amount of decision making allowed the parent or guardian be based on that individual's education and experience?

Several studies have shown that the use of informed consent has an impact on the results of research projects, often through a reduction in the spontaneity and naiveté of subjects. Investigating the effect of environmental noise, Gardner (1978) reported that performance aftereffects differed between subjects who received and signed an informed consent document and those subjects who were naive. Other studies (Dill et al. 1982; Resnick and Schwartz 1973) have shown that statements giving explicit permission to withdraw, even without any disclosure of hypothesis-related information, can significantly alter research findings. Do the results of these studies pose ethical issues concerning the use of informed consent? Should the potential for flawed research limit the use of informed consent in all or certain types of studies?

Hawthorne Effect

During the 1920s, the Hawthorne Works of the Western Electric Company was the site for a series of experiments investigating the effects of various working conditions on workers' performance (Roethlisberger and Dickson 1939). One variable, room lighting, was changed during a series of studies, and increased work performances were observed, even for the control group, producing findings contrary to what was expected. To investigate this phenomenon, the investigation introduced a "placebo control," which also produced changed work performance. Here, subjects were told they had been selected for participation in a study evaluating the effect of decreased illumination on work performance. The lightbulbs were regularly changed by maintenance workers in full view of the subjects. During this "placebo" phase of the study, the bulbs being removed were replaced with bulbs of equal power producing the same illumination, but the workers' performance decreased with each bulb change. The workers believed that each bulb change produced a decrease in lighting, although no actual decrease occurred. This effect, named the Hawthorne effect, has come to represent the idea that, in any program of intervention, variables not originally considered agents of change will bring about changes. Therefore, the Hawthorne, or placebo, effect refers to the finding that any scientific prestige suggestion may serve to increase or decrease the effectiveness of many programs of planned change or treatment.

One way researchers attempt to mitigate the Hawthorne effect is by using naive subjects. Naiveté can be accomplished through techniques such as using subjects who know nothing about the study or misleading subjects about the true nature of the study. This experimental method raises questions: Is it unethical to deliberately withhold information about the research from the subjects? Is deception of subjects concerning the nature of the study ethical? As indicated earlier, some data would be impossible to obtain without using naive subjects. This situation places responsibility and accountability on the researcher, requiring compromise and negotiation by both the researcher and the subjects. Should research requiring the researcher to withhold information from or deceive the subjects be banned? Does the potential ability of such studies to improve groups of people outweigh the possible harm, either mental or physical, an individual might experience through naiveté or deception?

According to basic DHHS policies established for the protection of human subjects, DHHS investigators are required to provide each subject with ''a statement that the study involves research, an explanation of the purposes of the research and the expected duration of the subject's participation, a description of the procedures to be followed, and identification of any procedures which are experimental'' (45 C.F.R. Subtitle A, §46.116(a)(1)). This requirement appears to preclude the conduct of any research that requires the use of naive subjects. Further, according to the language used, the use of deception and misleading subjects about the true nature of the study would be prohibited. Does this requirement for identification of procedures used in the research prohibit the use of placebo studies used to establish the efficacy of treatments without telling subjects which individuals receive the experimental drugs and which receive the placebo? If so, isn't the value of placebo controls eliminated and the resulting report therefore suspect?

In other research situations, scientists may misrepresent or disguise themselves so that subjects do not know they are being observed. Scientists may often assume other roles so that they are able to engage in undercover research. Thus, they may participate in the daily lives of those being studied, evaluating behavior through a technique called participant observation. One major advantage of this technique is that the subjects often either forget or don't know that researchers are observing their behavior and they reveal things they may not want recorded. Such scientists may be much more involved in the group dynamics than a passive observer would be. Consider the following case.

Case 3.2

The Participant Researcher

After becoming interested in the problem of illegal steroid use by teenage males, Dr. Knight Stealth decides to engage in undercover research as a participant observer. As a sport sociologist, he decides to assume a new identity and play a role in the underground sport drug culture. He believes this technique will help him obtain sociocultural information not otherwise available to him. To assume his role, he has a few oral placebo capsules made that resemble steroids, and he takes them in the presence of teenage steroid users while working out with them in gyms. His acceptance is complete, and some subjects come to him asking assistance in obtaining oral steroids like his. Dr. Stealth believes that if he refuses or gives the subjects one of his placebos, he will be exposed and that's the end of his study. Consequently, he obtains steroids and supplies them to his teenage male subjects and some of their friends. Acting as a sometime supplier, Dr. Stealth is able to study the group for one year before he drops out to write his research report.

Should Dr. Stealth give a full disclosure of the purpose of his study to the subject group? Should informed consent be secured from all individuals who will be observed during the study? Should such a study come under IRB review? How should an IRB handle its review of this study? Should the subjects be provided with a copy of the preliminary report and given the opportunity to delete any objectionable material before the results are released? Is it ethical that, as a participant observer, the scientist may have the opportunity to influence the behavior of the group being studied? Can a scientist be a participant observer and retain objectivity? What about ethical questions raised concerning the loss of privacy when subjects are observed by participant observers? Should investigators engage in illegal or immoral behaviors when functioning as a participant observer to gain acceptance by the group under study?

Using Animals in Research

Part of the ethical concern raised over the use of animal subjects is related to how the researchers view the animals. Recent reports (Arluke 1988) indicate that animals used in research projects are viewed as objects so that the humans associated with the research projects can psychologically distance themselves from the trauma of killing the animals during or after the project. This heightened concern by some has led to the formation of organizations, such as People for the Ethical Treatment of Animals, whose aim is to prevent cruelty to animals, eliminate the use of animals in research, and promote vegetarian diets, among other things.

Other ethical issues raised relate to the way animals are treated and motivated to perform. For example, many techniques are believed by animal rights groups to induce stress and fear. These would include using foot or tail shocks, withholding food, using heat and cold temperatures, and forcing animals to swim until they are exhausted. Considerable research describes the appropriate duration and intensity for treadmill running and other forms of exercise for humans, but how are such parameters decided when using animals as research subjects?

So, as with the use of human subjects, ethical issues are involved in research using animal subjects. What care should be afforded animal subjects? Is it appropriate to use computer models, in lieu of animal and/or human subjects, for research that will form the basis of policy decisions and treatments? Is the use of animals in research inhumane? What is a researcher's responsibility for the care and treatment of research animals? Perhaps underlying all issues relating to the use of animals is the debate summarized by Zelaznik (1993, 62):

> The debate concerning the use of animal subjects for research purposes provides a question without a correct answer. The issue is one of perspective. Is the life of an animal worth the same as or less than the life of humans? Does the value of an animal's life change as we move up or down the phylogenetic scale? Is a fruit fly worth less than a monkey or a golden retriever? These are questions that ethicists have grappled with for years. I surely do not have the answer. Putting aside the debate between animal researchers and antivivisectionists, the following questions can be framed. Given that in many animal studies the animal is killed as a part of the research process, what obligations do animal researchers have in conducting science so that the benefits to human and animal species outweigh the value of the life of the animal?

So, let's consider the ultimate question. Why use animals in research? Since other possibilities exist, such as tissue cultures and computer models, that might present the same types of information as can be gathered by using animal subjects, is it necessary to use animals as research subjects? Should experimentation using live animals in the pursuit of knowledge be restricted by ethical and social considerations? Must a balancing of priorities be made between the potential benefits that might accrue from live animal studies and the potential suffering that the study might cause to individual animals? If so, what constitutes suffering in species other than humans? Plants have hormones and enzymes and can adapt to surroundings. If a plant responds to pruning by oozing sap, is the plant suffering? Surely such a response indicates suffering of some sort. Cruelty exists in the wild when some animal species eat other species alive. Yet, should we extend human concepts of pain and suffering to animals or plants, and even if we should, should cruelty that causes suffering be classed as morally wrong? Why or why not?

Perhaps the answer lies not in the fact that cruelty to animals per se is morally wrong but that unnecessary cruelty to animals by humans represents a moral defect in humans and is consequently wrongful behavior. Another explanation may lie in the fact that many humans identify themselves with animals in such fashion that when animals suffer, they suffer, too, and so the only way to eliminate such human suffering is to remove the animal suffering. Yet, contradictory human behaviors exist that do not support this premise. Heroic medical efforts are often used in attempts to keep terminally ill humans alive, despite their suffering and expense, while an animal in similar circumstances will be "put to sleep" because it is deemed the humane thing to do.

So while the emotional attachment to animals exists with its attendant need to eliminate animal suffering, human dealings with animals are more complex than that. Our dealings with animals are based on the principle of utility and tempered with a principle that requires a duty of care so as not to engage in unnecessary exploitation of animals. Animals are used by humans in many ways, and they provide many benefits to humans. Some uses of animals, such as food production, result in the inevitable death of the animals; others, such as some scientific studies, police work, or sport, place animals at risk and involve some suffering. The counterbalancing principle is that of acting humanely, and inflicting minimum pain or injury serves to prevent unnecessary suffering by animals.

Animal research has produced many benefits for humans, including cures and/or treatment for diseases. To ban all use of animals in research would result in, or fail to alleviate, increased human suffering and death. Philosophically, most people believe that human welfare takes precedence over animal welfare. However, outspoken opponents of vivisection argue that all animal killing should be ceased and all humans should become vegetarians. They point to tissue culture and computer modeling as alternatives to the scientific use of animals. Others opposed to vivisection argue that only the necessary applied research, such as treatment of a particular disease, should involve animal subjects. The research projects that these groups wish to ban, however, will often produce results that do not have immediate practical uses but produce long-term benefits.

Animal research has long been an important part of biomedical research. Since 1896, the National Institutes of Health (NIH) has promoted the proper care and use of laboratory animals. The first guide for animal care was published by the Animal Care Panel (now the American Association for Laboratory Animal Science) in 1963. *The Guide for the Care and Use of Laboratory Animals*, now in its fifth edition (Committee on Care and Use of Laboratory Animals 1985), is the primary reference in the United States. The Pet Protection Act of 1966 was passed in response to an outcry that pets were being used in research studies. This act became the Animal Welfare Act (AWA) and had its latest revision in 1985. The act formalized Public Health Service (PHS) policy with a series of procedures required by investigators conducting research funded by governmental agencies. This act includes activities parallel to those associated with the use of human subjects. In 1973 the PHS required compliance with the AWA and the recommendations contained in the *Guide*. Today all vertebrates are covered under PHS policy, and institutions obtaining grants must obtain NIH approval before using animals in research. Therefore, each awardee institution has a responsibility for training staff for the management of their animal programs. See appendix E for some of the current rules and regulations governing animal research.

Matt (1993) proposed that the use of research animals is justified when there is no alternative. Animal research that is ethically justified appears to fall into five categories, according to Matt: drug testing, animal models of disease, basic research examining mechanisms not possible in human models, education of students with experience and knowledge gained from animal models, and the development of surgical techniques. These recommendations are similar to the current USDA guidelines given below (9 C.F.R. Chapter 1, Part 2. USDA Regulations: January 1, 1990).

Training and instruction of personnel must include guidance in several areas.

1. Humane methods of animal maintenance and experimentation, including

   [i] the basic needs of each species of animal;

   [ii] proper handling and care for the various species of animals used by the facility;

   [iii] proper pre- and postprocedural care of animals; and

   [iv] aseptic surgical methods and procedures.

2. The concept, availability, and use of research or testing methods which limit the use of animals or minimize animal distress.

3. Proper use of anesthetics, analgesics, and tranquilizers for any species of animal used by the facility.

4. Methods whereby deficiencies in animal care and treatment are reported, including deficiencies reported by any employee of the facility.

5. Utilization of services available to provide information on

   [i] appropriate methods of animal care and use;

   [ii] alternatives to the use of live animals in research;

   [iii] unintended and unnecessary duplication of research involving animals; and

   [iv] the intent and requirements of the Act.

According to the Public Health Service *Institutional Animal Care and Use Committee Guidebook* ([1992?], B-1), "federal regulations as well as generally accepted ethical principles" have two general goals: (1) to minimize scientific reliance on live animals and (2) to reduce pain, distress, and harm to laboratory animals to the minimum necessary to obtain valid scientific data. For example, federal legislation established the Institutional Animal Care and Use Committee (IACUC), whose purpose is to review research proposals involving animal subjects and develop standards of care for laboratory animals and animal facilities. When appropriate, nonanimal substitutes or a species of a lower phylogenetic order should be used. Grantee institutions must have an Institutional Animal Care and Use Committee in place to review research proposals involving animals as subjects and to oversee animal care and use.

As indicated earlier, the Helsinki Declaration of 1964 stated that biomedical research involving human subjects should be based on thorough animal experimentation. Toxic and carcinogenic substances are identified through research using animals as subjects. A vaccine that prevents HIV transmission in primates has been developed and is now being tested in humans. This research program emphasizes that the use of animals in biomedical research is as important as ever to scientists. However, the increased use of animals as subjects in scientific research and product testing has continued to raise concern over animal welfare, particularly in research that requires invasive techniques.

In addition to the development of policies seeking to reduce the number of animals used in research, alternatives to the use of animals for biological research and testing have been developed that may reduce the need for both human and animal subjects (Public Health Service [1992?], F-1):

> Alternatives to the use of animals include *in vitro* model systems. For example, isolated organ preparations such as the perfused heart represent *in vitro* models. More commonly, *in vitro* systems are thought of as cell, tissue and organ cultures. The source of the tissue may be primary tissue or cell lines, obtained invasively or non-invasively. Tests have been developed for using *in vitro* systems, for the evaluation of cytotoxicity, inflammation, genotoxicity, and target organ toxicity among others. A third alternative is the use of non-biological models systems. These include chemically based systems, physical models such as hemodynamic flow chambers or computer simulations. The quality of these systems is limited by the biological databases already available.

While these alternatives are pointed toward medical and biomedical research, it seems likely that exercise physiology and motor control research will be likely targets for the use of alternative models in the fields of exercise science and physical education.

A committee known as the Medical Research Modernization Committee has published materials highlighting types of research projects using animals that its members feel are inappropriate. *Perspectives on Animal Research*, edited by Kaufman and Todd (1989), takes the position that animal studies worked well in earlier biomedical research because that research centered on infectious diseases. In these studies, animals provided a place for the organisms to develop. The authors argue that animal research is not as appropriate with current disease studies for several reasons. Animals do not make appropriate subjects for studies attempting to understand mental and cognitive development in humans because the two nervous systems are not comparable and the human mental experience and brain biochemistry are distinctly different. The use of animals in some cancer research is also criticized because humans and animals may respond differently to the same agents. Metabolisms, anatomical structures, and immune systems differ and animals synthesize different biochemical elements. Consequently, toxins will react differently in animals and humans. In studies of vision, animals are not appropriate subjects, as cats don't possess a macula or fovea and rabbits differ in eyelid construction, tearing mechanisms, and cornea structures. Therefore, according to the Medical Research Modernization Committee, the types of investigations conducted today are not conducive to the use of animal subjects.

Another recent development in biomedical research has been the development of computer modeling that can be used to replicate physiological responses in humans and animals. Computer modeling is simply the use of mathematics, usually algebraic equations, which allows the researcher to manipulate variables and see what should happen to the subject. A model is merely a simplified replication of the real thing, whether an animal or human subject. Consequently, in science, the mathematical model is the use of mathematical languages rather than ordinary languages to represent a situation that exists in the real world. The advantage of a mathematical model is that it provides quantitative conclusions and is extremely reproducible. Because the mathematical processes being used are simple, the mathematical models can be run on modern personal computers and the software is readily available. Therefore, today the process is much more readily available to researchers than earlier when the lack of or limited capacity of personal computers required the use of a mainframe computer.

Two problems are present in the use of computer modeling or mathematical models that do not exist when using human or animal subjects. First, it is just as easy to make an error when developing a mathematical model as it is for the researcher to make an error when using ordinary language. However, finding errors in mathematical models is often more difficult than finding them in ordinary language. The second problem lies in the apparent precision provided by a mathematical model. Data can be analyzed and the results carried out to an infinite number of decimal places, which can entice the reader into believing that the whole procedure is far more precise than is the case.

For a moment, consider the possibilities associated with computer modeling. The researcher will not be faced with defending the project before an IRB. There will not be problems with securing subjects, subjects missing data, and controlling for extraneous factors. The investigator will not be faced with an inadequate supply of data, because the computer can be told to generate as much data as necessary. The investigator will not have to deal with missing or incomplete data. On the other hand, what kind of problems will computer modeling produce? Should data obtained from tissue culture and computer modeling studies be confirmed using animal studies? Consider the scenario suggested in the following case study.

Case 3.3

The Computer as a Research Subject

Dr. Cynthia Cipher is an exercise physiologist who spent her post-doctoral fellowship in a medical college working with a biomedical researcher who made extensive use of mathematical models during research projects. Subsequently, Dr. Cipher obtained her own software and began to apply it to the type of research projects that she conducted as a professor at Smart University. She adapted the mathematical models to her research and began to input values from her past research to see if the models provided results that were representative of the results she obtained using human subjects. When these results approximated the values obtained with subjects, she began to input data that was representative of subjects' responses and used her software to predict experimental results. Once she obtained these results, she wrote up the data in typical experimental style and submitted manuscripts to professional journals. She described her subjects, methods, and analyses as if she had tested human subjects and never revealed that she was using mathematical models instead. Dr. Cipher found that this process greatly decreased the time required to conduct research and publish it in professional journals.

Does the manner in which Dr. Cipher described her subjects and research procedures pose any ethical concerns? Should journals limit their acceptances to investigations using human or animal subjects? What implications would follow if every researcher decided to use computer modeling rather than human or animal subjects? Should Dr. Cipher publish her software and research technique, or should it be considered a "trade secret"? Other issues associated with a switch to tissue culture and computer modeling studies follow.

A major limitation to restricting research to tissue cultures and computer models lies in the fact that some drugs and treatments have effects that are not localized to the tissue involved in the test. They may produce side effects that lead to the discovery of other uses. Such is the case with Lithium, which is prescribed for the treatment of manic-depression as well as cancer, migraine headaches, and overactive thyroid gland, or Nortriptyline, which is prescribed for depression but which can also be used to control migraine headaches. Penicillin was discovered by accident, resulting from a contaminated sample. Such discoveries would be greatly reduced or nonexistent if research were limited to the use of computer models and tissue cultures. Likewise, basic work on protein synthesis using animal subjects is not classified as applied research producing an immediate application. Consider, however, that current research investigating protein synthesis may lead to many important applications in the field of reproduction, hormone synthesis, antibodies production, and treatment or prevention of diseases such as HIV and AIDS that are caused by viral agents.

Thus, while the use of tissue cultures and computer modeling would eliminate some ethical considerations regarding the use of animal subjects and would be easier for researchers to use, there remains information that can only be obtained by using animal subjects. The simplicity and focus of tissue culture studies are both their strength and downfall. The use of muscle tissue culture will enable the researcher to determine what the effect of a treatment will be on muscle tissue, but any general physiological or behavioral effects that may result will be unknown without the use of animal experiments. In the future, tissue culture and computer modeling studies will become more important and widely used, but the need for animal experimentation will remain to determine final safety and efficacy of a given treatment.

Data Collection

One problem area in research often cited by the NIH and others is that of data collection and storage. According to NIH policy, failure by the "grantee" to provide primary (original) data triggers the presumption that the data does not exist. Several institutions that are heavily funded by governmental agencies issue a bound notebook with pages numbered in sequence to the investigator in charge of a funded project, and all data must be placed in this book. Once the research projects are completed, the laboratory notebooks are stored in secured facilities so that they can be produced upon demand. This process might solve the problem of providing primary data upon request, but it does not successfully prevent all unethical conduct associated with data collection. In particular, these policies will not prevent fabrication and falsification of data or recording and logging problems.

The scientific method requires a systematic, controlled investigation. Thus, the null hypothesis is developed and the variables that relate to the study of that hypothesis are selected. Data collection and statistical analysis must be developed to answer the question posed in the null hypothesis. Must a proposal and study be limited to an evaluation of the null hypotheses posed, or can an investigator include other variables that may provide information for later analyses that do not relate to the question in the present proposal? How does one select the variables that will be included or excluded from the study?

Quantitative Data

Some research projects involve such a volume or rapid input of information that manual recording of data is impossible. Data is entered directly into a computer, and electronic storage represents the original or primary data. In other cases, the samples (e.g., some blood or tissue assays) require analysis using expensive equipment or specialized training not possessed by the investigator, so the data is generated by commercial laboratories. Situations may exist in which the data may be retained by another entity (e.g., large databanks stored on electronic media). In each case, "original, handwritten" records are not possessed by the grantee. Does the use of new technology create an ethical problem? What responsibility does the investigator have for records recorded through electronic media? By commercial laboratories? For storage of data held by other entities? Does the researcher have an ethical duty to "prove" that the data is accurate and not falsified? If so, how should this duty be met?

Qualitative Data

Researchers who engage in investigations, such as historical studies, that use qualitative data often feel they do not need to concern themselves with ethical issues. After all, they are not using human or animal subjects or collecting numbers that can be massaged. They do not need to concern themselves with the appropriateness of statistical techniques. If these researchers think that they are exempt from ethical concerns because they engage in qualitative data studies, they are only deceiving themselves. Every research study has the potential for abuse. Consider the discussions concerning "political correctness," or PC. It is well known that certain groups, such as women, ethnic populations, and some religious bodies, did not always have the opportunities that they enjoy today. Should a historical researcher color the past situations to bring the data in line with current thinking? Should unfavorable findings be ignored and left out of the report? Should practices that were engaged in during the past be taken out of social and cultural contexts?

Good historical research practices require that the researcher acquire data from primary sources whenever possible. Often, primary sources or pertinent information may arise from citations placed in other papers or publications. To check the information for accuracy, the investigator should review the primary source. A temptation exists to use the secondary information while citing the primary source if that resource would be difficult or expensive to obtain. As with all research, "*Truth* begets *truth* and *error* begets *error*" (Massey 1973, 40). While such researchers do not collect the same kind of data as those engaging in laboratory research, they are charged with the same goals: eliminate bias, be intellectually honest, and keep their research free from their opinions and prejudices. They are taught to engage in both internal and external criticism to ensure the information is as reliable and valid as possible. A temptation to skip procedures to facilitate publication is present in this research as in quantitative studies. All researchers, regardless of their use of qualitative or quantitative data, must be sensitive to ethical concerns they face when designing studies, collecting data, analyzing data, and reporting their results.

Data Analysis

Ethical questions will arise during the process in which the researcher determines the manner in which the data will be analyzed and what procedures will be used to provide the quantitative information that will be subjected to statistical

Table 3.1 Analytic Methods Available for Urine Drug Testing

Method	Abbreviation
Complete specific and highly sensitive	
Gas chromatography–mass spectroscopy	(GC-MS)
High specific and sensitive	
Gas chromatography	(GC)
High-pressure liquid chromatography	(HPLC)
Low specificity and variable sensitivity	
Enzyme-multiplied immunoassay	(EMIA)
Radioimmunoassay	(RIA)
Variable specificity and poor sensitivity; nonquantitative	
Thin-layer chromatography	(TLC)

treatments. Consider, for example, that drug use by high school athletes is being investigated. Once the urine specimen has been collected, several methods can be used to detect the presence of prohibited substances (see table 3.1). The reliability of these methods varies considerably, as does the cost. The most accurate and reliable analytic method (GC-MS) is also the most expensive, while the least accurate and reliable method (TLC) is also the least expensive. The latter method not only will produce false negatives but will also produce false positive results.

Should a researcher propose or continue a study, knowing that the method used will not give the most accurate results? While this decision may not impact subjects if complete confidentiality is provided and no individual information is released, it will affect the results. Should the investigator assume that the false positive and false negative results are random occurrences and therefore cancel an experimental bias and use that technique to save money? Should the researcher run a pilot study comparing the most reliable and least reliable techniques to see if false results are random and if the above assumption is true? Should the investigator decide not to take any chances and to use the most accurate analytical method? When reporting the data, should the investigator discuss the reliability and accuracy problems inherent in the analysis procedures used with urine specimens?

Statistical Treatment

Statistical treatments are also subject to abuse when inappropriate techniques are used for analysis and research results are improperly interpreted. Computers have made techniques possible that would have been discarded in the past because of the time and expense involved. For instance, some researchers have been reported to collect their data and analyze it using a large variety of statistical techniques. Only after all the analyses have been reviewed will they decide which results to report using the post hoc selection of the analyses to provide the desired results. This process is contradictory to the requirements imposed by the scientific method whereby all procedures and analyses that will be followed are determined prior to initiation of the research. This is a fundamental procedure used to eliminate investigator bias. Consequently, researchers need to seek advice prior to beginning an investigation so that they can select analytical procedures based on the methodological issues existing in the project. Further, all publication and reporting of research findings must contain information describing the research design, data collection, and methods used for the analysis of results in sufficient detail for others to understand, replicate, and evaluate the report.

Summary

All researchers face ethical choices, whether they engage in quantitative or qualitative studies. It makes no difference whether the data comes from a research laboratory or the field. Ethical questions will arise during an investigation. Great concern has been voiced about the ethical use of human and animal subjects in research. Perhaps this concern has overshadowed ethical issues that exist in other parts of the study, such as the practice of manufacturing and ''cooking'' data to produce the desired results. A balance needs to exist between the desire to protect human and animal subjects and the need for researchers to investigate populations that will provide data required for policy decisions. Honesty and integrity are important characteristics for a researcher to possess. These characteristics require that the investigator consider the ethical concerns that arise during all phases of the research.

CHAPTER 4

HANDLING YOUR RESULTS

All of the work completed during the formation of the problem and the development of the research methods has been directed toward producing results that can be tested against the hypothesis proposed. Once the data has been collected and analyzed, the product of the study becomes visible. At this point the investigator is faced with several new ethical questions. Chief among these is how to ensure that the participants' privacy is not compromised. The investigator must also decide how to present the findings so that new knowledge is presented clearly, accurately, and without misrepresentation. The issue of confidentiality is a continuation of earlier ethical questions surrounding the use of human subjects in research.

Confidentiality

Research, especially that conducted for federal agencies, often involves the collection of potentially sensitive information from individual subjects. Such information might pertain to economic and social status, home and family relationships, physical and mental health, or participation in illegal activities, such as illicit drug use. Confidentiality has become even more critical as the improved technology used to obtain, store, and use information has been coupled with the increased acquisition of potentially harmful information. Most notable among the technological improvements affecting privacy has been the widespread installation of computers and communication equipment, including networks, that enable researchers to communicate worldwide. These technological changes have facilitated both the accumulation and storage of information and the ability to access sensitive information. A growing number of firms provide public information electronically. In spite of these developments, concerns about confidentiality and accessibility of information in the United States are not new.

Beginning shortly after the Constitution was adopted in 1787, the collection of census information has been used to meet the constitutional requirement for determining the composition of the House of Representatives. Recognizing early concerns about confidentiality, in 1840 the U.S. Census Bureau developed guidelines that required the census enumerators to regard the information they collected as confidential. Social and educational research increasingly exposed potential problems associated with a lack of confidentiality that existed concerning the plentiful data collected about individuals, so Congress enacted several laws and regulations designed to protect individual confidentiality. The privacy of students was addressed by the Privacy Act of 1974 and the Family Educational Rights and Privacy Act of 1974. The Freedom of Information Act and other more specific statutes have been enacted to tighten the confidentiality further. Notable among these are 12 U.S.C. 8, 9 prohibiting disclosure of census records; 42 U.S.C. 242m regulating release of data collected by the National Center for Health Statistics; and 42 U.S.C. 4582 and 21 U.S.C. 1175 protecting patient records connected with drug abuse programs and research activities connected with federal agencies or departments.

Taking a contrary approach, many state laws mandate that teachers, nurses, lawyers, doctors, and social workers report all actual or suspected child abuse. According to these laws, any professional who meets children who may have been abused is obligated to report the incident and to keep any evidence available. What is the ethical duty of investigators who have told subjects that their data will be kept confidential—and based on these assurances received approval from parents for their children to participate—if they detect a suspicious injury to a subject? Consider the following example of the discovery of possible child abuse by an investigator and the issues of duty owed to different parties.

CASE 4.1

TO REPORT OR NOT REPORT SUSPECTED CHILD ABUSE

A researcher developed a growth and development study that included pubescent assessment that would enable him to see the child when nude. This was a longitudinal study that would involve the subjects for more than ten years. All subjects and their parents had been assured that all data would be kept confidential, no data about the individual would be reported, and that the results would be reported in such a manner that no child or school could be identified. The question of confidentiality had arisen during a meeting between the researcher and parents, and he assured the parents that all phases of the study and any observations would be kept confidential. During the assessments and the data collection the researcher noticed bruises and other signs of possible child abuse. The state law mandates that he report any actual or suspected cases of child abuse to the County

Child Study Bureau. The researcher did not detect any broken bones or apparently serious physical injuries.

Does the researcher's assurance of confidentiality to the parents include a promise not to report possible child abuse? Does the seriousness of the injury affect the researcher's duty to report child abuse? The researcher knows that if he reports suspected child abuse, it will affect this subject pool and the research. Since he has collected data for three years and has seven more years of data collection to complete, he is concerned about the future of the study if he reports his suspicions. Do his responsibilities lie with the research or with state law? What are his responsibilities to the subject(s)? Does your answer change depending on how the study is being funded? If the researcher has a contract with an outside agency for financial support, what duty does the researcher have to that agency to complete the study? If the researcher reports his suspicion of child abuse and the case goes to trial, what role, if any, should the researcher play as a witness? Should the researcher refuse to testify about his observations?

Investigations (Boruch 1975; Singer 1978) have shown that response rates of research subjects will vary with the nature of the assurance of confidentiality, the strategies used to ensure confidentiality, and the sensitivity of the information requested. Subjects who do not understand or trust the procedures used to produce confidentiality are less likely to provide sensitive information. Even in face-to-face interviews involving the exchange of sensitive information, assurances of confidentiality produced higher response rates from subjects.

In spite of researchers' promises to preserve subjects' anonymity, some situations prevent total privacy. For example, some case studies may divulge enough information about subjects that readers may guess at their identities. Researchers may contract to study organizational or other aspects of a company or a community. Even when they employ fictitious names, their reports may provide enough information so that knowledgeable individuals can identify the company or community involved. Further, the data may often be embarrassing or detrimental to the subjects. This is a problem frequently faced by researchers who rely on the case study method for their investigations, as described in Case 4.2.

CASE 4.2

CONFIDENTIALITY ISSUES IN A CASE STUDY APPROACH TO RESEARCH

In addition to teaching and conducting research, a professor who specialized in sport psychology also worked as a consultant for collegiate and professional athletic teams. This consultant business gave her access to many well-known college and professional athletes. In a class lecture concerning the benefits provided by sport psychologists, she used a case study of an athlete as a teaching tool and gave many details of problems involving an athlete with whom she recently had worked. The details included conflicts, legal problems, illegal behavior, and other incidents affecting the athlete's performance. She also described their impact on the athlete's performance during the season. Because some incidents had received widespread publicity, many of her students could identify the individual she was describing in the case study.

What expectation does a famous athlete or other well-known person have with regard to confidentiality and privacy? If the person is not well known, do the expectations for privacy change? Does your answer depend on whether one or many people can identify the subject? Does the presentation of this athlete's need for a sport psychologist represent a breach in confidentiality and an invasion of privacy, or is it fair to use details of his life in an educational setting? Is it ethical for her to use a real athlete as a class example, or should she have created a composite, hypothetical athlete? If she uses a hypothetical athlete, should it have any similarity to real athletes who might be identified? Consider the same situation, but instead of the sport psychologist presenting private details about an athlete in a class lecture, she publishes them in a professional journal. Is this unethical? Should the investigator refrain from publishing case studies because individuals might be identified from the data presented? Might this problem be resolved by using an IRB and informed consent? Even if the athlete should provide informed consent to use the data, does that permit the researcher to release data that might enable some readers to identify the individual?

In situations where promises of anonymity will not provide adequate confidentiality, statistical techniques have been developed to safeguard data. These techniques are used with increasing frequency in situations where the deletion of direct personal identifiers will not adequately preserve confidentiality. Among the techniques used are *microaggregation methods*, which substitute synthetic average persons for individual data; *randomized response methods*, in which interviewers record the answers to randomly determined questions preventing others from determining which question a specific subject is actually answering; and *error inoculation methods*, in which the data is "tainted" in some way by the insertion of random errors.

Thus, either providing or failing to provide confidentiality may raise ethical questions. How much assurance of confidentiality must a researcher provide? If the data shows that illegal activities are being conducted, should a researcher risk jail to maintain confidentiality? Does the use of error inoculation in research raise ethical concerns about the data? Must the investigator describe the error inoculation process, since one requirement of good science is that the study is reproducible? If the experiment is described so that it is reproducible, does that destroy confidentiality? Must absolute confidentiality be provided in all cases?

If absolute confidentiality is not always to be provided, what determines how much confidentiality must be provided? Is it ethical for a researcher to state he or she is providing confidentiality of response yet code questionnaires so the identity of those who have and have not responded can be determined? Does potential embarrassment represent enough harm to require absolute confidentiality by the researcher? If so, what constitutes damaging information and who determines it? The researcher? The subject? The IRB?

In many institutions, computer accounts and storage are available to faculty, researchers, and students on the institution's mainframe computer. In other cases, these individuals may store data on the hard drive of a personal computer in their office or laboratory. Individuals who use a central computer are required to use a password to protect their data, and they should change this password regularly to prevent unauthorized use of their account. Is it ethical for a faculty member to give students access to a central computer account that provides access to data that may require confidentiality? Is it ethical for a faculty member to use a password that is easily remembered or bypassed such as their name or birth date? Does storage of data on a personal computer hard drive or floppy disk without the use of security such as passwords raise any ethical questions regarding privacy? Does viewing confidential data stored on a computer by anyone other than the researcher represent "publication" of the data? What type of care and how much care are owed subjects with regard to confidentiality?

Reporting Results

Until recently, the method of choice used to report the results of a study was publication in a refereed journal, preparation of a report to the granting agency, or publication of larger works. With the earlier mentioned pressure to publish and to obtain grants to support research, researchers have announced their findings in additional ways, most commonly through press conferences or news releases. The reason for this change in mode of information dissemination is an attempt to gain any advantage over competing researchers for the limited funds available.

Problems With Early Release of Findings

An example of potential abuse that exists when one uses this process occurred in 1989 when investigators at the University of Utah called a press conference to announce that they were able to achieve nuclear fusion at room temperature. They chose this method of reporting findings rather than waiting to have their findings successfully complete the peer review process associated with publication in professional journals. Their claims were controversial, and subsequent investigations failed to replicate their findings. It was not certain that the original researchers even took the time to replicate their original study. After several years, controversy still existed as to whether these investigators did achieve the results they reported in their press conference. Certainly a rapid presentation of their results was in their interest, especially to aid in attempts to obtain funding for their continuing efforts. Or perhaps they chose this mechanism because they believed they would face extreme difficulty in getting such an unusual finding approved in a peer review process. Whatever their reasoning, a press conference for reporting findings proved counterproductive for these researchers.

As mentioned earlier, AIDS research is completed with a heightened sense of urgency. One reason given for this is the pressure to *do something* so that people can escape the ravages of a fatal disease. Not only has the pressure led researchers to ignore time-honored procedures associated with the scientific method but it has prompted early release of findings, many of which are only tentative. Another reason for the early release of AIDS research results is financial. AIDS research news releases are used because the frontrunner gets more money, whether it is an advantage in obtaining grants or getting a patent process on new drugs having potentially beneficial effects. Several years after the change in government policy was effected to allow release of preliminary data under an accelerated approval plan, concerns were raised about the appropriateness of this change (Gorman 1994). Several AIDS treatment activists have called for a return to the more thorough drug testing process, even if it means a delay in receiving approval for new treatments. Does the early release of information, through a press conference, involve ethical decision making? Should research findings be distributed through press releases prior to publication in professional refereed journals? What role, if any, should news releases play in the research process?

Misrepresentation of Findings

Several techniques can be used to misrepresent the findings of a research investigation. In his book *How to Lie With Statistics*, Huff (1954) provides an entertaining and informative discussion of statistical frauds. For example, the scientific method requires that generalization of findings be limited to groups that the subject pool represents. However, many reports outside peer-reviewed manuscripts will decline to identify or describe the population from which the subjects are derived, leaving the reader to guess at the amount of generalization justified by the results. Worse, many reports will suggest that the findings apply to all. Findings that fail to support the research hypothesis or prior research that provides contrary results may be omitted.

We are bombarded with such claims in the popular press. "This product is 50% better." However, we must ask ourselves, better than what? In what way? How was this determined? "More doctors recommend this product." What kind of doctors? How many doctors were involved in the study? Why do they recommend it? Does it make any difference? In other instances, inappropriate statistics are used in the report. For example, means might be reported when the median is a more meaningful and representative statistic. Coincidental findings might be reported as crucial. Reliability or variability statistics might be omitted so that only measures of central tendency are provided in the report. Nonsignificant differences might be described as showing trends. What ethical decisions must be made when developing the research report? Does overgeneralization of findings represent an inappropriate ethical choice?

Graphs and figures are commonly distorted to misrepresent findings of investigations. Widths of bars can vary, violating the convention that the bars in a graph are of equal widths. Differences in coloring or shading will violate the convention that the bars in a graph must be constructed in like manner. The relative proportion convention of the X axis and Y axis can be violated, producing problems in scale. Pictograms are the most easily distorted of all graphic methods; here an increase of two times in height and two times in width will produce a fourfold increase in total area, greatly distorting a reported twofold increase. Inappropriate scales might be used when constructing a graph, or part of an axis might be omitted without the reader being told. When an axis does not contain data continuously to zero, a break in the axis may be omitted so the reader has difficulty in determining the true situation. Does the choice in graphic presentation present an ethical problem? Does the caveat "reader beware" apply to research reports? Should it? Does the author's decision to display the findings in the "best light" present an ethical problem?

Undoubtedly, good editorial work will serve to eliminate manuscripts that contain this kind of distortion, but there remains a necessity to consider what responsibility the author of the manuscript has with regard to presentations. If the researcher writes an article for a popular rather than scientific journal, does he or she have the same or a different duty regarding presentation of the findings? In other words, does the sophistication of the reviewers or readers change the author's responsibility? If so, what determines that responsibility—the commercial need to sell copies, or the scientific duty to present unambiguous, accurate information?

Questions Remaining at End of Study

Several questions remain after the results of the study have been obtained and the investigator has set about securing publication of the manuscript. One question revolves around the safekeeping of the data. Institutional IRBs will require that the data be kept for a specified period and will request the location of the stored data. But who owns the data: the investigator who conducted the study or the institution that provided the support? Who should be included as a coauthor, and who is not recognized as such? The answers are not always as simple as they may appear.

Other institutions may not have policies that demarcate procedures as clearly as the federal government. Often universities have written policies that apply to patents and other types of work product but not to manuscripts. But an issue remains unresolved and unstated. If a faculty member uses university facilities, money, and time to conduct research or write a book, who should receive any financial benefit that results? Should the person who performed the intellectual work receive all the benefit, or should the institution that made the intellectual work possible receive the financial profit? If a faculty member writes a book and keeps the royalties, should he or she also receive financial merit awards for the effort? Or should the institution receive the royalties and the faculty member only receive merit pay for the work? Presently most institutions ignore this issue, but as funding continues to tighten, these matters will surely receive more consideration.

Publication

The final part of the scientific process is the publication of the study's results. Publication accomplishes two things: it makes the new knowledge available to others, and it allows for replication and reproduction of the study. To accomplish these goals, the publication must be complete and provide necessary information about how the study was conducted. The authors of the study must have contributed to its completion, they must support the conclusions, and they must be ready to respond to inquiries.

Authorship

Authorship is the means by which researchers receive credit for work they have completed and their work is made available to the scientific community for review and replication. In today's academic climate, authorship in peer-reviewed journals is a requirement for tenure and promotion. The knowledge explosion has produced increased specialization and the situation where perhaps 80 percent of those who have conducted research are still alive and 50 percent of the publications have occurred in the past ten years. Consequently, multiauthored or collaborative studies have increased, with each author being responsible for a part of the manuscript. As a result of this practice, few coauthors understand, or are able to defend, the entire study and its general conclusions. Often authors who are not statisticians have little or no understanding of the statistical analyses used or the computer programs that provide the analysis. Perhaps the first question to be answered is, who should be included as an author for a manuscript? Consider the following situation, for example.

CASE 4.3

THE DEPARTMENT CHAIR AS A PHANTOM AUTHOR

As a department chair at Intellect University, Dr. Symbiotic has little time to conduct his own research, let alone write for publication.

To remedy this problem, Dr. Symbiotic requires that all faculty members in his department attach his name to their publications or presentations as a coauthor. After all, the reason he is lacking the time to conduct his own scholarly research activity is the departmental work he does on the faculty members' behalf. Dr. Symbiotic reasons this practice is due compensation for that loss.

Does it make sense that the department chair is included as a coauthor on all manuscripts published by his department faculty members? What if the department chair is placed as a coauthor in recognition of financial support he provided for the faculty members from department funds? Would your answer change if the coauthor were director of a research laboratory where the study was conducted instead of the department chair? How much should an individual have contributed to a study before he or she is listed as a coauthor?

Are all researchers who are listed as authors responsible for understanding the general conclusions and defending the study? Does the definition of "fraudulent science" include the practice of including the head of a laboratory as the lead author on all publications coming out of the laboratory? Is this practice one of misrepresentation when the laboratory head listed as a coauthor has little or no role in the research being published? If a graduate student publishes research and has received extensive assistance from the faculty advisor and/or others, what authorship recognition should be provided for their assistance in the research?

The Experimenter's Role

Over the years, research has become more complicated and a team approach to research has become more prevalent. This change in the way research is conducted has resulted from the need for increased technical competencies, increased specialization in knowledge areas, and increased size and complexity of the research projects. Often research projects will include the principal investigator who is responsible for the subject area, a researcher specializing in the technical analysis who is responsible for data development, and a statistician who is responsible for data analysis. The result is that large, complicated projects may exceed the competency of any one individual scientist. Ethical questions arise when publishing the results of cooperative research ventures. What is the role of a researcher, whether sole investigator or part of a team, during publication of the experimental results? In other words, who is responsible for the veracity of the publication? Does the responsibility lie only with the principal investigator? Are all members of the research team equally responsible for the accuracy of the publication? If the team consists of an exercise physiologist and a statistician, are both team members responsible for the entire publication?

It is not only faculty members who are subjected to the pressure to publish or perish. In today's highly competitive market for faculty appointments and grant support, students are also often faced with the need to have completed research and publications before obtaining a graduate assistantship or a professional appointment. Often students may experience difficulties in getting their manuscripts published because of lack of experience in writing or because of lack of notoriety. The following is a case that was litigated over the methods which a student chose to have a manuscript published.

CASE 4.4

IN SEARCH OF A PUBLISHED ARTICLE

A doctoral student knew that his professional advancement depended on publication in refereed journals. So he submitted a manuscript reporting his master's thesis research to a leading journal in his field. After his manuscript was rejected, the student decided to add his advisor's name to the manuscript as a coauthor, as his advisor was a well-known authority in the field. He did not tell his advisor of this decision, and the advisor never received or reviewed a copy of the manuscript. After the advisor's name was added and the manuscript was resubmitted, the article was published in a prestigious journal. Only after publication did the advisor and his employer discover that he had been credited as a coauthor (*Slaughter v. Brigham Young University*, 514 F. 2d 622 [10th Cir. 1975]).

Does publication of the article using the advisor's name to provide inferred expertise, approval, and support of the work by an authority represent scientific misconduct and fraud on the part of the graduate student? What statement does this case make about the ethics of a journal that would not publish research submitted by an unknown professional but that readily accepted the same study after an authority's name was added as coauthor? What if the advisor had not reported the student to university authorities but had added the publication to his own vita? Would that represent an ethical problem on the part of the advisor?

Summary

Research often involves the collection of potentially sensitive data, such as economic information, medical and health status, or legal problems. Improvements in technology have made both the collection and dissemination of data easier and more accessible. Concern over privacy continues to be voiced by individuals and agencies, even though Congress has passed laws that address these needs. Not only has Congress been concerned, but researchers have also questioned the need for confidentiality assurances given to potential subjects. Studies have shown that assurances of confidentiality result in higher response rates than when no such assurances are present. Statistical techniques have been developed for use when anonymity will not provide adequate confidentiality. However, while protecting privacy, these techniques will also greatly affect the investigators' ability to reproduce the study, one of the key parts of the scientific method.

In the quest for recognition and funds, many researchers first release the results of their investigations through sources other than scientific journals. Such release may open the door for misrepresentation in an attempt to present their research in the "best light" in hopes that the publicity will enhance their chances for increased funding. The potential to present data in many different forms can create ethical problems for investigators, especially when they are writing for popular publications rather than scientific journals. Authors should carefully consider the ethical choices that exist when developing the format through which they will present the results of their investigations.

CHAPTER 5

PREVENTING FRAUD AND MISCONDUCT

Since fraud, dishonesty, and other forms of scientific misconduct are known to be present, the approach usually adopted to deal with these problems has focused on the detection and punishment of wrongdoers. This approach requires the use of due process and a procedure to provide the accused with opportunities to rebut the charges. If a researcher is found in violation of professional standards, a variety of sanctions are imposed. These sanctions fall into two categories: those imposed internally by the university or institution employing the researcher and those imposed by an external source such as governmental funding agencies. The types of sanctions frequently imposed are listed in table 5.1.

Sanctions may or may not be coupled with rehabilitation because not all persons who violate ethical principles or act without integrity are amenable to rehabilitation. Unfortunately, there is no way to predict who will or will not be rehabilitated. One's character will determine whether misconduct occurs, and character is rather resistant to change. Repeat offenders are likely because of the pressure society places on success. If repeat offenders are to be eliminated, the present motivational system with its attendant pressures must be changed. At the very least, proper mentoring is required, including adequate contextual patterns and mechanisms.

According to the Final Rule of the Public Health Service, published in the *Federal Register* August 8, 1989, intent is not required for a finding of scientific misconduct. Therefore, mistakes can be treated as misconduct. Researchers active in funded research, however, recommend a bifurcated approach that differentiates between misconduct and mistakes. This approach would produce differences in the disciplinary handling of the individual who is found to have engaged in misconduct as opposed to having been guilty of scientific mistakes. It has been proposed that sanctions should be imposed on individuals found guilty of scientific misconduct; for individuals found guilty of committing scientific mistakes, remedial activities are proposed, such as oversight and monitoring.

One reality facing the scientific and academic community is that not all researchers who engage in scientific misconduct will benefit from, or be amenable

Table 5.1 Types of Sanctions Imposed for Fraud in Research

Internally imposed sanctions

1. Verbal reprimand
2. Letters of reprimand
    a. Not a part of the permanent record
    b. Included in the permanent record
3. Monitor research with prior review of all publications
4. Supervision of grant submissions
5. Salary freeze
6. Promotion freeze
7. Restriction of academic duties
8. Termination of work on the project
9. Reduction in professorial rank
10. Separation from university without loss of benefits
11. Separation from university with loss of benefits
12. Fines to cover costs

Externally imposed sanctions

1. Revocation of prior publications
2. Letters to offended parties
3. Discontinuance of service to outside agencies
4. Prohibition of obtaining outside grants
5. Release of information to agencies, professionals, newspapers, and others
6. Referral to legal system for further actions
7. Fines to cover overhead costs

to, rehabilitation. Unfortunately, there is no way to predict who can or cannot be rehabilitated. An individual's character will determine whether he or she will engage in misconduct. Since character is a stable part of an individual's makeup, professional admonitions and policies will not produce a change in character. This means that any change in conduct on the part of some individuals will require a change in the motivational system and the pressure placed on the individual. Proper mentoring, coupled with adequate content patterns and mechanics, is one technique that can be used to produce a change in behavior. Individuals who are pressured to succeed are likely to be repeat offenders. Better results in behavioral change come when the context in which the individuals work is modified, for example, working under close supervision by senior colleagues. Still, the principal deterrent to fraud in science and research will be the high probability that fraudulent data will be detected upon its release and publication.

In response to the need for scientific integrity, several professional organizations and academic institutions have developed policies and procedures directed toward reducing academic and scientific dishonesty. However, as suggested earlier, the prevention of fraud and misconduct must be the responsibility of each member of the scientific and academic community. If research is to have integrity, each researcher must take responsibility for the honesty and accuracy of his or her project. Faculties must adhere to and teach high ethical standards as well as research techniques and academic competencies.

The situation in Case 5.1 involves a faculty member who published her own textbook based on questionable research that consisted of anecdotal summaries. Further, she required students to purchase this text for use in her course.

CASE 5.1

ANECDOTAL RESEARCH FOR PUBLICATIONS

After several years of practice as a sport psychologist, Dr. Carla Caricatura decided to write a sport psychology textbook. During her practice, she developed a self-improvement program for athletes that included techniques of mental practice, relaxation, and focus or concentration activities. Based on her experiences she developed a program, Mentis Intellectus Vis Viva, that is sold as a videotape, audiotape, and book. She collected anecdotal, rather than empirical, research that she cites as unpublished case studies, unpublished manuscripts, and correspondence in her textbook. A large portion of the textbook is based on the undocumented reports that support the use of Mentis Intellectus Vis Viva as a sports psychology technique. She never mentions that she is the publisher of Mentis Intellectus Vis Viva or that she generated all the unpublished case studies, unpublished manuscripts, raw data, and correspondence. Dr. Caricatura also requires all her students to purchase the sport psychology textbook.

Does the use of anecdotal, rather than empirical, research in a textbook raise ethical concerns? If the book's preface stated that the contents were anecdotal tales so the course could be taught as a case study, would there be ethical concerns? Does the practice where professors require students in their classes to purchase books they wrote raise ethical concerns? Should a disclosure of all the facts, or lack of facts, be made in these books? If this behavior is unethical, what can be done to prevent its occurrence?

Rules and guidelines can be developed to cover procedural issues such as data storage, presentation of the proposed study to an IRB for approval, announcement and publication of research findings, and authorship of manuscripts. However, elimination of outright dishonesty in research such as manufacturing data, plagiarism, and breaking blinding codes cannot be prevented by a written code of ethics, as these problems reflect the character of the researcher involved. Changing the character of the researcher and preventing such an unethical individual from conducting research are the methods required to prevent such fraudulent practices from occurring. Research has not indicated that either practice is currently possible.

Since developing a code of conduct that applies to all possible research situations is difficult, if not impossible, it is critical that an informative approach be used during the training of scientists, rather than relying solely on a regulatory code of rules. Following such a scenario, ethical decision making as well as the expectations of professional and governmental agencies are taught. The purpose is to teach researchers how to make ethical decisions. Such decisions will be based on the values and experiences of others engaged in similar research or on an appeal to some authority.

No one agrees on what constitutes an authority that can be used in research decision making. Some recognize religious principles, while others look for authority outside religion. Common appeals to authority outside religion for use in ethical decision making might include the following:

- The individual is precious, and the individual's benefit takes precedence over the society.
- Equality is of the utmost importance, and everyone must be treated equally.
- Fairness is the overriding guide to ethics, and all decisions must be based on fairness.
- The welfare of society takes precedence over that of the individual, and all must be done for the benefit of society.
- Truth, defined as being genuine and conforming to reality, is the basis for decision making.

Each of these appeals to authority requires that the researcher understand the consequences of its use. Some of the issues and questions raised through the use of these authorities are as follows. What is the result when all is accomplished for the benefit of single individuals rather than the collective good of society? We recognize that all individuals possess equal rights in our system of government, but not all individuals have equal abilities or needs; therefore, equal treatment may produce unequal, disparate results. The concept of fairness is also fraught with difficulties; what is fair for one individual may not be fair for another. What happens to an individual when the society takes precedence over the individual? Should society be viewed as more important than individuals? Is it important that an individual have rights? What is truth? Is truth a series of facts, or is truth a fundamental knowledge or belief? How do you know the truth? Ethical decision making requires a careful balancing of many issues, needs, and rights.

Many scientists, according to Reece and Siegal (1986), support the ideal of attaining knowledge for its own sake. Individuals following this approach believe that knowledge is intrinsically valuable, that is, valuable in itself and not because it can produce other ends. Thus, our culture upholds education, knowledge, and understanding. We extol the virtues of education and pride ourselves on being civilized. This emphasis on the pursuit of knowledge leads to moral issues and systems of ethics, particularly in terms of the authority used for decision making.

Ideal utilitarians hold that truth is the sole authority that should be used to guide ethical decision making in science. They see truth as an independent human good that ought morally to be sought for itself, not just for the satisfaction that occurs during the pursuit. Following this approach, the authority is based on the belief that "truth and beauty are ideal values and that morality consists in promoting these values as well as pleasures" (Reece and Siegal 1986, 61). Truth, according to *Webster's Dictionary*, is "the body of real things, events, and facts; a judgment, proposition, or idea that is true or accepted as true; agreement with fact or among true facts or propositions." However, the question asked earlier remains. Is truth the facts, the data, or is truth the understanding and ideas that flow from the analysis of data? If truth is the understanding or ideas, how do we know what is the truth when reasonable people's ideas can disagree about data analyses? If truth is the facts or data, how can truth exist without meaning?

An unethical investigator can still circumvent this approach to maintaining scientific integrity. Thus, if any improvement in ethical conduct and integrity within the scientific community is to occur, a shift in emphasis and increased efforts toward the prevention of problems must be coupled with the deterrent factor of early detection and punishment of violators. The data retention rule required by federal agencies and adopted by most academic communities is perhaps the most critical ingredient in prevention of misconduct. This requirement must be coupled with the need for researchers to follow the scientific method when developing and conducting research. An important part of the scientific process is distribution of knowledge. The part of this process used to promote integrity and honesty in science and research is publication of manuscripts in professional journals based on peer review.

Refereed Publications

One factor leading to improved integrity is the high probability that scientific dishonesty will be detected. Refereed publications are one mechanism through which detection can occur. Professional journals require that the manuscripts they publish are accurate and honest. Articles published in professional journals can have an impact on agency policy decisions and treatments used in medicine or during therapy and must, therefore, present accurate and replicable information. Hence, the selection of manuscripts to be published relies on the peer review process. Peers evaluate content, writing quality, and the scientific process.

The type of activity described in Case 5.2 is difficult to control unless a reviewer receives the same manuscript to evaluate from two or more journals.

Case 5.2

The Issue of Multiple Submissions

Dr. Ima Fitz is a new assistant professor at Mensa University on the tenure track. Recognizing a need to obtain several publications, she submits identical manuscripts to several journals simultaneously for review. When the manuscript is accepted by a journal, she formally requests that the other journals return the manuscripts to her for additional modifications, or she does not return copyedited manuscripts or page proofs and asks that the manuscript not be published by them. She feels the simultaneous submission practice will speed up her publications, promotion, and acquisition of tenure and will improve her acceptance rate by journals. She regularly signs statements that a journal has received her sole submission, even though several copies are floating around at other journals.

Are multiple submissions of the same manuscript unethical, or is it only unethical when the author allows the same manuscript to be published in more than one journal? Does her willingness to lie about the matter raise ethical concerns? If Dr. Fitz's colleagues discover what she is doing, should she be disciplined? If so, how should she be disciplined?

When a manuscript is accepted for review, several peers evaluate it to see if it relates to the scope and mission of the journal. Since most professions have several areas of specialization, or areas of focus, a single individual is unlikely to be knowledgeable in every area. Thus, reviewers who are representative of the target population can determine the manuscript's appropriateness.

Publication of a book or journal requires that all the chapters or articles be written in a uniform style. Thus, manuals of style have been adopted. Reviewers assist the editor when they evaluate whether the manuscript being considered meets the style requirements and contains good grammar and sentence structure. Poorly written manuscripts do not convey scientific information effectively.

Perhaps most important, peer review helps promote integrity in science. Reviewers who are knowledgeable in a field of expertise can evaluate the scientific method used in the study. They can determine whether the analysis adequately tests the hypothesis formed to evaluate the problem. The methods, subjects, and protocol can be scrutinized to see if they were appropriate to test the problem.

Methods are checked to see if sufficient detail is present to allow other investigators to repeat the study. Data presented in the tables and results can be examined to see if any errors or inconsistencies are apparent. Citations are checked for accuracy, and the discussion is evaluated for linkages between the current findings and prior published reports. Once the scientific method aspects of the study have passed scrutiny and the manuscript is published, reasonable expectation exists that the process used by the author was appropriate and that the study has integrity. Publication in a refereed journal means that the study has been presented in sufficient detail that it is open and others can repeat the process and provide corroboration. The veracity of the information has been checked through independent review by others and has met the criteria for publication.

Institutional Responsibility

The question of how to prevent unethical behavior in science and research has been approached in several ways. Federal agencies have adopted rules that grantees must obey when receiving funds (copies of these rules are contained in the appendixes). Professional organizations have formulated guidelines and recommendations, some of which have become policy, in attempts to heighten members' awareness of the need for ethical behavior. One of the leaders in this area has been the Association of American Medical Colleges. In June 1982 this association published ten recommendations directed at member institutions. While the following guidelines form the basis for the development of institutional policies concerning ethical science and research in medical colleges, all research institutions should consider them when making policies:

- "Create a climate that promotes faithful attention to high ethical standards."
- Have a conspicuous and understandable mechanism in place for dealing with instances of alleged fraud.
- Adopt institutional policies that define misrepresentation of research data as a major breach of contract between the individual and the institution.
- Articulate institutional policies that foster openness of research.
- Encourage faculties to discuss research ethics to heighten awareness and recognition of these issues.
- Establish policies to (1) provide an appropriate and clearly defined locus of responsibility for conduct of research, (2) offer assurances that individuals charged with the supervision of other researchers can realistically execute their responsibility, and (3) pay particular attention to adequate supervision of large research teams.
- Assure that quality rather than quantity of research is emphasized as the criterion for the promotion of faculty.
- Examine institutional policies on authorship of papers and abstracts to ensure that the named authors have had a genuine role in the research and accept responsibility for the quality of the work being reported.

- Review policies on the recording and retention of research data to ensure that the policies are appropriate and clearly understood and that all faculty comply.
- Examine the institutional role and policies in guiding faculty concerning public announcement and publication of research findings.

The National Institutes of Health (1990) has developed guidelines for the conduct of research that can be adopted by institutions with little or no modification. The guidelines for data management, publication practices, and authorship are particularly appropriate for use by institutions. The guidelines are published in pamphlet form, and the pertinent parts of this pamphlet are highlighted in the following paragraphs.

Data Management

One integral part of the scientific process is that the experiment must be reproducible. Management and conservation of the primary data are one mechanism through which replication can occur. If the data is produced through instrumentation or external laboratories, the acquisition of the data must be detailed. All data should be recorded in a form that will allow access, complete with annotations and indexing necessary for review. In collaborative research, all participants should know the status of all data used, have access to the data, and be able to defend the data.

Investigations funded by contract will usually specify who owns the data. In the case of NIH-funded studies, all data is owned by NIH, but the rights of publication rest with the investigator. The responsibility for providing the data rests with the grantee, which according to NIH policy is the institution. The institution can delegate custodial responsibility for the data, but must always be able to provide the original data for review. The principal investigator is entitled to a copy of the data, rather than the original data.

Often the data is contained in a laboratory notebook, but any data recorded by electronic means must also be stored in original form as well as any other information, hard copy, formula, program, or substance required by the project. After publication, the research data and any unique requirements for analysis, such as reagents or computer programs, should be made available to any responsible scientist seeking further information. The original data must be maintained for the required period of time to allow analysis and repetition by others. Usually this period is from five to seven years, but it can vary with the circumstances.

The NIH position has been supported by a U.S. Supreme Court decision that held when a governmental agency having oversight for a grant exercises supervisory activities, but does not take permanent custody of the data, the raw data remains the property of the organization doing the research. A summary of the court's position on this ownership question is provided in the *Forsham v. Harris*, 445 U.S. 169; 63 L. Ed. 2d 293; 100 Sup. Ct. 978 (1980) case.

This legal action resulted after a privately controlled organization of physicians and scientists specializing in the treatment of diabetes studied the effectiveness of certain diabetes treatment regimens supported by grants from the National Institute of Arthritis, Metabolism, and Digestive Diseases (NIAMDD). Under pertinent federal regulations governing the grants, NIAMDD exercised some supervision by reviewing periodic reports and conducting on-site visits; however, the day-to-day administration was left to the grantee. Further, although NIAMDD had a right of access to the data collected by the grantee pursuant to the grants and could obtain permanent custody of the grantee's data, NIAMDD did not exercise its right to review or to obtain custody of the data. Ultimately, the grantee's reports indicated that the use of certain drugs for the treatment of diabetes increased the risk of heart disease. This led to proceedings by the DHEW and the Food and Drug Administration (FDA) to control labeling and use of the drugs. This action prompted a national association of physicians, critical of the grantee's study, to request access to the raw data. After both the grantee and DHEW had denied the association's requests for access, the association brought an action in the U.S. District Court for the District of Columbia to require DHEW to make available all the raw data compiled by the grantee under the federally funded study. The District Court refused to order access to the raw data, holding that DHEW had properly denied the physicians' request. Subsequently, the U.S. Court of Appeals for the District of Columbia Circuit affirmed (190 App DC 231, 587 F2d 1128).

This decision was further appealed to U.S. Supreme Court, which affirmed the lower court decision. It was held that the written data that was generated, owned, and possessed by the privately controlled organization receiving federal funds from an agency subject to the Freedom of Information Act did not constitute "agency records" within the meaning of the enforcement provisions of the act, and that the data was not "agency records" merely because the grantee was subject to some supervision by the agency in the use of federal funds which the grantee received.

Publication Practices

Each publication should make a unique and substantial contribution to its field rather than be one of several fragmentary publications of the work. Further, the publication should be timely and not part of multiple publications of the same or similar data. The publication should contain sufficient information to enable an informed reader to assess its validity and to replicate the results. Thus, the manuscript must be complete enough that it provides all information that would be necessary for other peers to replicate the experiment. Unique materials and programs that are essential for repetition of the published work must be made available to other qualified researchers.

Authorship

Authorship refers to the credit given through a listing of names of the study participants. It is the realization of the scientist's responsibility to communicate research results. Authorship is also the mechanism through which the allocation of credit for scientific advances and an assessment of a scientist's contributions to developing new knowledge are effected. The privilege of authorship should be based on a significant contribution to the completion of the research and is based on conceptualization, design, execution, and/or interpretation of the study. Individuals who contributed through advice, reagents, analyses, space, financial support, etc., should be acknowledged but not listed as authors. All questions of authorship should be discussed and resolved before and during the course of study. In the case of multiple authors, each author should fully review all material that is to be submitted for publication. Each author should be willing to support the general conclusions of the study and be willing to defend it. The submitting author should be considered the primary author having the additional responsibility of coordinating the completion and submission of the work, satisfying the rules of submission, and coordinating responses to inquiries. All collaborators should have reviewed the manuscript and have authorized submission of the manuscript. An illustration about the types of problems that may arise with coauthorship is presented in Case 5.3.

CASE 5.3

AN UNWILLING COAUTHOR

Three graduate students complete a research project for an exercise science class, submit their paper to the professor, and obtain an excellent grade for the work. Two of the students decide to submit the class project for publication in a national journal, but they don't tell the third student until after they've mailed it. The third student tells the others that the work should not be submitted because it has a couple of design flaws; for example, there is no reliability information provided in the study. The others reply that it's too late to argue. When the third student asks that his name be taken off the manuscript, the others refuse, saying, "You may not need the publication, but we do, and your name will help us get it published. Your past publication record will help us." They add, "If the study is no good, the reviewers will not approve it for publication." Nothing further is done by any of the authors. The first two authors receive the proofs, make corrections, and say nothing to the third author. The study is eventually published, still lacking reliability data, and the third author finds out the manuscript has been published (with his name as a coauthor) when reading the journal.

What ethical questions are raised about the students who wish to use another's name to assist in getting the manuscript published? What should the student do if he does not want his name to be used? How can this situation be avoided?

Another example of how authorship credit for published material can be a problem is presented in the example shown in Case 5.4. Consider the ethical questions that arise when someone uses material generated by others without giving credit where credit is due. It is unlikely that a publisher would accept a manuscript that had a multitude of authors for each chapter as would be the case in this situation.

CASE 5.4

A SHORTCUT TO AUTHORSHIP

Dr. Purloin requires all his students to submit term papers on assigned topics. He keeps copies of these papers. After the courses are completed, he extracts, word for word, parts of the papers that are appropriate to his area of expertise and combines them into a textbook, which he then submits to a publisher under his own name. Dr. Purloin does not change much, if any, of the students' work. He only adds transitions so that the material flows together coherently. Dr. Purloin keeps all book royalties and submits this book for his promotion and merit evaluations. Dr. Purloin is listed as the sole author, and he does not acknowledge the students' contributions.

Who has the responsibility for preventing such a situation? Suppose a student or another professional suspected what had happened? What would be their ethical duty? If they have a duty to report their suspicions, whom should they inform? The author? The publisher? The author's institution? The professional association?

Individual Responsibility

Perhaps one of the more challenging questions concerns the researcher's personal behavior. As a researcher, are you held responsible for your conduct in areas that involve personal choice? For example, many people drink socially. Should there be a requirement that a researcher must be held to a standard of responsible alcohol consumption? It is widely known that chemically dependent individuals cause turmoil in their family, work, and other relationships. Other individuals assume the role of co-dependent, acting as people-pleasers, martyrs, or stoics to protect or work around the chemically dependent person. Others may believe that the problem will go away if they wait long enough, or they don't want to be considered a "snitch" or "stool-pigeon." Is a professional responsible for trying to prevent problems normally associated with substance abuse by co-workers?

Today federal and state laws require equal employment opportunity, non-discrimination, and harassment-free workplaces. Today most states have laws that deal with child abuse. These laws generally define child abuse and categorize it as physical, sexual, and mental or emotional, impose duties on professionals to report abuse, set penalties for those found guilty of abuse, and establish the administrative procedures that are followed in abuse cases. Similar laws relate to discrimination and sexual harassment in the workplace. Does a researcher have an ethical responsibility to know and follow the law, including the reporting of individuals who violate its provisions, or is the researcher a detached scientist who is responsible only for the conduct of well-designed, unbiased scientific studies?

Proposing a universal solution to the fraud that occurs in research, or academe in general, is fraught with problems. Kretchmar (1993) observed that rule ethics have generally been more popular than situational ethics, and he suggested that this situation may result because the moral aspects of different ethical dilemmas are in fact similar. However, he warns that rule ethics bring their own potential problems to the solution of ethical problems. For example, rules may be too specific for all situations, or, conversely, the rules may be too vague. Rules may contradict one another, or rules may become more important than the purpose for which the rules were generated, that is, the welfare of society. The development of useful ethical rules to aid in resolving ethical problems in science might be a difficult task, but it should not be viewed as impossible.

Perhaps the most useful way for researchers to deal with ethical problems faced in scientific investigations is to use an axiom ethical theory, which is analogous to the axiom systems used by science that satisfy the criteria of consistency, universality, relative completeness, and parsimony. Looking at axioms as maxims that are widely accepted on their intrinsic merit or as a proposition regarded as a self-evident truth and a maxim as a general truth, fundamental principle, or rule of conduct, philosophers can begin to build a system to aid ethical decision making and conduct. Such a system might begin with a general proposition followed by a series of principles and axioms based on logic and deduction. A system applicable to research might be developed in a manner as displayed in the following brief example.

PROPOSITION: Science should be honest, ethical, and based on integrity.

Principle 1: Scientific investigations should be conducted according to the scientific method.

Axiom 1.1: All research should be conducted in accord with the use of null hypotheses.

Axiom 1.2: All subjects participating in a study should be representative of the population under investigation.

Axiom 1.3: All data should be collected in an accurate, appropriate manner.

Principle 2: Scientific reports should be accurate, complete, and verifiable.

Axiom 2.1: All authors should be responsible for understanding and defending the study.

Axiom 2.2: Only individuals who have made significant contributions to the investigation should be listed as authors.

Axiom 2.3: All coauthors should have given their permission to be listed as such.

Using this type of approach to the solution of ethical dilemmas, it's possible to develop a workable system to guide decision making. In addition, such a system will call attention to problems that exist in formation, conduct, and dissemination of research. However, rules by and of themselves will not eliminate the possibility of fraud in science.

The maintenance of integrity and honesty in research and science requires more than governmental and institutional policies, regulations, and guidelines. It requires appropriate behavior on the part of each individual researcher. It seems fitting to end with the following suggestions that each scientist can adopt to become more sensitive to appropriate and ethical behavior in research.

Therefore, as a researcher, I will

- follow the scientific method, using the built-in checks and balances that eliminate bias in research;
- prepare academically and technically to the extent necessary to conduct the research in my field;
- look for ethical issues that exist when conducting research and use appropriate decision-making practices to resolve them;
- be knowledgeable of and implement governmental practice and guidelines in research where appropriate;
- be aware of the need for honesty in science and research, supporting its practice in my own work and in that of my peers;
- not plagiarize or engage in other dishonest behavior such as trimming, cooking, faking, or stealing data and ideas;
- support and defend the general conclusions of any study for which I am author or coauthor;
- keep research data so it is immediately available to scientific collaborators and supervisors for review; and
- be knowledgeable of and conform to the standards of professional behavior expressed through codes of ethics.

Summary

In closing, a large number of the problems observed in research can be eliminated by following strict protocol during the planning of the study, selection of subjects, collection of data, and writing the results. However, protocol alone will not completely remove misconduct from the research process. Dishonest individuals who feel a need to "fudge" or "cook" or make up data will do so given the opportunity. Two approaches might be used to minimize this problem: (1) do everything possible to prevent the existence of situations where researchers can make up data; and (2) through peer monitoring and institutional policy, eliminate dishonest scientists from the professions. Now is the time when we must invest the time and energy into research to make every effort for honesty.

QUESTIONS FOR DISCUSSION

The following questions are included to provide additional topics for class discussion over and above those contained in the text. For each question, the ethical issues should be identified and suggestions developed to eliminate the problems. Consider the development of rules that would be controlling in the situation and how effective such rules would be. Further consider where the responsibility for each problem lies and, if any disciplinary action should be taken, what it should be.

1. A well-known researcher is mentor to a graduate student who helped with a number of experiments. The student elects to replicate some of the researcher's earlier research and develop an independent publication record. This study is more compact and only looks at a few of the variables used in the previous study. The results of the student's study contradict the earlier published and widely accepted works. What should the researcher do?

2. A university professor receives a research grant from Great Performances, Inc., which markets a ''high-energy'' drink supplement promoted as a performance enhancer. The professor knows that if their product is successful, and if the research supports their claims, he will receive more funding for future research. One way to achieve this goal is to include this product as a variable in all of his studies that evaluate athletic performance, regardless of the original purpose of the study. What is the professor's responsibility? How should he proceed?

3. A university researcher applies for a grant to obtain funds to study sports drinks as a means of maintaining longer aerobic work. The researcher sits on the board of the company that will supply the sports drinks used in the study. Does this situation present a conflict of interest?

4. The health education professor applies for a grant to study the effects of alcohol on health. The results of this study will be used to determine government policy toward regulation of the alcohol industry. Through investments, the professor owns three hundred shares of a company that produces alcoholic beverages, among other things. This is a small, highly profitable company, and its alcoholic

products have contributed substantially to its success. Does the professor's ownership of stock in the company present a conflict of interest? If not, would ownership of a significant number of shares pose an ethical problem? Does ownership in any form or amount pose an ethical problem?

5. A university professor is involved in research for a sponsoring agency that provides significant funds for the project. Without notifying either the university or sponsoring agency, the researcher attempts to orient the research to meet the agency's needs, rather than following the traditional approach to scientific research. Does the practice of tailoring one's research to the needs of the funding agency represent misconduct? Should the professor tell the university and/or agency what she is doing? Should a colleague be a ''whistle-blower'' and inform the university or agency?

6. A biomedical health scientist owns part interest in a clinic through investments. The scientist applies for research grants and writes in funds to pay part of the clinic overhead as necessary research expenses. Salaries listed in the grant proposal include money to pay students who staff the clinic. The students are described as research assistants. However, they will not participate in the research project directly, although they will occasionally meet with some of the research subjects when they come to the clinic for therapy. Does this behavior on the part of the students represent ethical scientific activity?

7. An exercise physiologist has included blood chemistry work in a study supported by NIH. Lacking the laboratory equipment and skills necessary for analysis, the physiologist has farmed out the analyses to a clinic owned by his brother. The selection of the clinic was made solely by the physiologist, and he solicited no competitive bids before or after the award. Does this subcontract represent a conflict of interest? Is the subcontract unethical?

8. A researcher is involved in government-sponsored projects. A federal agency also employs her as a consultant. Does this dual capacity present an ethical problem? Does it make a difference whether the consultantship is with the same agency that is sponsoring the research or with a different agency?

9. A professor leaves the academic setting to accept a government position. After a few years, he returns to the university. He then prepares grant proposals, which he submits to the government agency where he once worked. Many of the people with whom he worked are still at the government agency. Is this ethical?

10. A faculty member who served as a corporate consultant was listed on the corporation's letterhead in such a manner that his university employer could be perceived as supporting corporate activities and practices. Does this violate any ethical behaviors?

11. During a study of the effectiveness of NCAA enforcement practices, a researcher obtained information regarding athletes' use of drugs and other prohibited activities. Athletic departments and coaches who had recruited the players wanted access to the data so they could institute educational programs and other appropriate practices aimed at preventing such behaviors in the future. Failure to turn over the information would likely prevent future access to athletes for studies conducted by this researcher. What should the researcher do?

APPENDIX A

THE NUREMBERG CODE

The Proof as to War Crimes and Crimes Against Humanity

Judged by any standard of proof the record clearly shows the commission of war crimes and crimes against humanity substantially as alleged in counts two and three of the indictment. Beginning with the outbreak of World War II, criminal medical experiments on non-German nationals, both prisoners of war and civilians, including Jews and ''asocial'' persons, were carried out on a large scale in Germany and the occupied countries. These experiments were not the isolated and casual acts of individual doctors and scientists working solely on their own responsibility, but were the product of coordinated policy-making and planning at high governmental, military, and Nazi Party levels, conducted as an integral part of the total war effort. They were ordered, sanctioned, permitted, or approved by persons in positions of authority who under all principles of law were under the duty to know about things and to take steps to terminate or prevent them.

Permissible Medical Experiments

The great weight of evidence before us is to the effect that certain types of medical experiments on human beings, when kept within reasonably well-defined bounds, conform to the ethics of the medical profession generally. The protagonists of the practice of human experimentation justify their views on the basis that such experiments yield results for the good of society that are unprocurable by other methods or means of study. All agree, however, that certain basic principles must be observed in order to satisfy moral, ethical and legal concepts:

1. The voluntary consent of the human subject is absolutely essential.

This means that the person involved should have legal capacity to give consent; should be so situated as to be able to exercise free power of choice, without the intervention of any element of force, fraud, deceit, duress, overreaching, or other ulterior form of constraint or coercion; and should have sufficient knowledge and comprehension of the elements of the subject matter involved as to enable him to make an understanding and enlightened decision. This latter element requires that before the acceptance of an affirmative decision by the experimental subject there should be made known to him the nature, duration, and purpose of the experiment; the method and means by which it is to be conducted; all inconveniences and hazards reasonably to be expected; and the effects upon his health or person which may possibly come from his participation in the experiment.

The duty and responsibility for ascertaining the quality of the consent rests upon each individual who initiates, directs or engages in the experiment. It is a personal duty and responsibility which may not be delegated to another with impunity.

2. The experiment should be such as to yield fruitful results for the good of society, unprocurable by other methods or means of study, and not random and unnecessary in nature.

3. The experiment should be so designed and based on the results of animal experimentation and a knowledge of the natural history of the disease or other problem under study that the anticipated results will justify the performance of the experiment.

4. The experiment should be so conducted as to avoid all unnecessary physical and mental suffering and injury.

5. No experiment should be conducted where there is an *a priori* reason to believe that death or disabling injury will occur; except, perhaps, in those experiments where the experimental physicians also serve as subjects.

6. The degree of risk to be taken should never exceed that determined by the humanitarian importance of the problem to be solved by the experiment.

7. Proper preparations should be made and adequate facilities provided to protect the experimental subject against even remote possibilities of injury, disability, or death.

8. The experiment should be conducted only by scientifically qualified persons. The highest degree of skill and care should be required through all stages of the experiment of those who conduct or engage in the experiment.

9. During the course of the experiment the human subject should be at liberty to bring the experiment to an end if he has reached the physical or mental state where continuation of the experiment seems to him to be impossible.

10. During the course of the experiment the scientist in charge must be prepared to terminate the experiment at any stage, if he has probable cause to believe, in the exercise of the good faith, superior skill and careful judgment required of him, that a continuation of the experiment is likely to result in injury, disability, or death to the experimental subject.

APPENDIX B

World Medical Association

DECLARATION OF HELSINKI

Recommendations Guiding Medical Doctors in Biomedical Research Involving Human Subjects

Adopted by the 18th World Medical Assembly, Helsinki, Finland, 1964, and as revised by the 29th World Medical Assembly, Tokyo, Japan, 1975.

Introduction

It is the mission of the medical doctor to safeguard the health of the people. His or her knowledge and conscience are dedicated to the fulfillment of this mission.

The Declaration of Geneva of the World Medical Association binds the doctor with the words, "The health of my patient will be my first consideration," and the International Code of Medical Ethics declares that, "Any act or advice which could weaken physical or mental resistance of a human being may be used only in his interest."

The purpose of biomedical research involving human subjects must be to improve diagnostic, therapeutic and prophylactic procedures and the understanding of the aetiology and pathogenesis of disease.

In current medical practice most diagnostic, therapeutic or prophylactic procedures involve hazards. This applies a fortiori to biomedical research.

Medical progress is based on research which ultimately must rest in part on experimentation involving human subjects.

In the field of biomedical research a fundamental distinction must be recognized between medical research in which the aim is essentially diagnostic or therapeutic for a patient, and medical research, the essential object of which is purely scientific and without direct diagnostic or therapeutic value to the person subjected to the research.

Special caution must be exercised in the conduct of research which may affect the environment, and the welfare of animals used for research must be respected.

Because it is essential that the results of laboratory experiments be applied to human beings to further scientific knowledge and to help suffering humanity, the World Medical Association has prepared the following recommendations as a guide to every doctor in biomedical research involving human subjects. They should be kept under review in the future. It must be stressed that the standards as drafted are only a guide to physicians all over the world. Doctors are not relieved from criminal, civil and ethical responsibilities under the laws of their own countries.

I. Basic Principles

1. Biomedical research involving human subjects must conform to generally accepted scientific principles and should be based on adequately performed laboratory and animal experimentation and on a thorough knowledge of the scientific literature.
2. The design and performance of each experimental procedure involving human subjects should be clearly formulated in an experimental protocol which should be transmitted to a specially appointed independent committee for consideration, comment and guidance.
3. Biomedical research involving human subjects should be conducted only by scientifically qualified persons and under the supervision of a clinically competent medical person. The responsibility for the human subject must always rest with a medically qualified person and never rest on the subject of the research, even though the subject has given his or her consent.
4. Biomedical research involving human subjects cannot legitimately be carried out unless the importance of the objective is in proportion to the inherent risk to the subject.
5. Every biomedical research project involving human subjects should be preceded by careful assessment of predictable risks in comparison with foreseeable benefits to the subject or to others. Concern for the interests of the subject must always prevail over the interest of science and society.
6. The right of the research subject to safeguard his or her integrity must always be respected. Every precaution should be taken to respect the privacy of the subject and to minimize the impact of the study on the subject's physical and mental integrity and on the personality of the subject.
7. Doctors should abstain from engaging in research projects involving human subjects unless they are satisfied that the hazards involved are

believed to be predictable. Doctors should cease any investigation if the hazards are found to outweigh the potential benefits.

8. In publication of the results of his or her research, the doctor is obliged to preserve the accuracy of the results. Reports of experimentation not in accordance with the principles laid down in this Declaration should not be accepted for publication.
9. In any research on human beings, each potential subject must be adequately informed of the aims, methods, anticipated benefits and potential hazards of the study and the discomfort it may entail. He or she should be informed that he or she is at liberty to abstain from participation in the study and that he or she is free to withdraw his or her consent to participation at any time. The doctor should then obtain the subject's freely given informed consent, preferably in writing.
10. When obtaining informed consent for the research project the doctor should be particularly cautious if the subject is in a dependent relationship to him or her or may consent under duress. In that case the informed consent should be obtained by a doctor who is not engaged in the investigation and who is completely independent of this official relationship.
11. In case of legal incompetence, informed consent should be obtained from the legal guardian in accordance with national legislation. Where physical or mental incapacity makes it impossible to obtain informed consent, or when the subject is a minor, permission from the responsible relative replaces that of the subject in accordance with national legislation.
12. The research protocol should always contain a statement of the ethical considerations involved and should indicate that the principles enunciated in the present Declaration are complied with.

II. Medical Research Combined With Professional Care (Clinical Research)

1. In the treatment of the sick person, the doctor must be free to use a new diagnostic and therapeutic measure, if in his or her judgment it offers hope of saving life, reestablishing health or alleviating suffering.
2. The potential benefits, hazards and discomfort of a new method should be weighed against the advantages of the best current diagnostic and therapeutic methods.
3. In any medical study, every patient—including those of a control group, if any—should be assured of the best proven diagnostic and therapeutic method.
4. The refusal of the patient to participate in a study must never interfere with the doctor-patient relationship.
5. If the doctor considers it essential not to obtain informed consent, the specific reasons for this proposal should be stated in the experimental protocol for transmission to the independent committee (I, 2).
6. The doctor can combine medical research with professional care, the objective being the acquisition of new medical knowledge, only to the

extent that medical research is justified by its potential diagnostic or therapeutic value for the patient.

III. Nontherapeutic Biomedical Research Involving Human Subjects (Non-clinical Biomedical Research)

1. In the purely scientific application of medical research carried out on a human being, it is the duty of the doctor to remain the protector of the life and health of that person on whom biomedical research is being carried out.
2. The subjects should be volunteers—either healthy persons or patients for whom the experimental design is not related to the patient's illness.
3. The investigator or the investigating team should discontinue the research if in his/her or their judgment it may, if continued, be harmful to the individual.
4. In research on man, the interest of science and society should never take precedence over considerations related to the well-being of the subject.

APPENDIX C

42 CODE OF FEDERAL REGULATIONS, CHAPTER 1

§ 50.103 Assurance—Responsibilities of PHS Awardee and Applicant Institutions

(a) *Assurances.* Each institution that applies for or receives assistance under the Act for any project or program which involves the conduct of biomedical or behavioral research must have an assurance satisfactory to the Secretary that the applicant:

(1) Has established an administrative process, that meets the requirements of this subpart, for reviewing, investigating, and reporting allegations of misconduct in science in connection with PHS-sponsored biomedical and behavioral research conducted at the applicant institution or sponsored by the applicant; and

(2) Will comply with its own administrative process and the requirements of this subpart.

(b) *Annual Submission.* An applicant or recipient institution shall make an annual submission to the OSI as follows:

(1) The institution's assurance shall be submitted to the OSI, on a form prescribed by the Secretary, as soon as possible after November 8, 1989, but no later than January 1, 1990, and updated annually thereafter on a date specified by OSI. Copies of the form may be requested through the Director, OSI.

(2) An institution shall submit, along with its annual assurance, such aggregate information on allegations, inquiries, and investigations as the Secretary may prescribe.

(c) *General Criteria.* In general, an applicant institution will be considered to be in compliance with its assurance if it:

(1) Establishes, keeps current, and upon request provides the OSIR, the OSI, and other authorized Departmental officials the policies and procedures required by this subpart.

(2) Informs its scientific and administrative staff of the policies and procedures and the importance of compliance with those policies and procedures.

(3) Takes immediate and appropriate action as soon as misconduct on the part of employees or persons within the organization's control is suspected or alleged.

(4) Informs, in accordance with this subpart, and cooperates with the OSI with regard to each investigation of possible misconduct.

(d) *Inquiries, Investigations, and Reporting—Specific Requirements.* Each applicant's policies and procedures must provide for:

(1) Inquiring immediately into an allegation or other evidence of possible misconduct. An inquiry must be completed within 60 calendar days of its initiation unless circumstances clearly warrant a longer period. A written report shall be prepared that states what evidence was reviewed, summarizes relevant interviews, and includes the conclusions of the inquiry. The individual(s) against whom the allegation was made shall be given a copy of the report of inquiry. If they comment on that report, their comments may be made part of the record. If the inquiry takes longer than 60 days to complete, the record of the inquiry shall include documentation of the reasons for exceeding the 60-day period.

(2) Protecting, to the maximum extent possible, the privacy of those who in good faith report apparent misconduct.

(3) Affording the affected individual(s) confidential treatment to the maximum extent possible, a prompt and thorough investigation and an opportunity to comment on allegations and findings of the inquiry and/or the investigation.

(4) Notifying the Director, OSI, in accordance with § 50.104(a) when, on the basis of the initial inquiry, the institution determines that an investigation is warranted, or prior to the decision to initiate an investigation if the conditions listed in § 50.104(b) exist.

(5) Notifying the OSI within 24 hours of obtaining any reasonable indication of possible criminal violations, so that the OSI may then immediately notify the Department's Office of Inspector General.

(6) Maintaining sufficiently detailed documentation of inquiries to permit a later assessment of the reasons for determining that an investigation was not warranted, if necessary. Such records shall be maintained in a secure manner for a period of at least three years after the termination of the inquiry, and shall, upon request, be provided to authorized HHS personnel.

(7) Undertaking an investigation within 30 days of the completion of the inquiry, if findings from that inquiry provide sufficient basis for conducting an investigation. The investigation normally will include examination of all documentation, including but not necessarily limited to relevant research data and proposals, publications, correspondence, and memoranda of telephone calls. Whenever possible, interviews should be conducted of all individuals involved either in making the allegation or against whom the allegation is made, as well as other individuals who might have information regarding key aspects of the allegations; complete summaries of these interviews should be prepared, provided to the interviewed party for comment or revision, and included as part of the investigatory file.

(8) Securing necessary and appropriate expertise to carry out a thorough and authoritative evaluation of the relevant evidence in any inquiry or investigation.

(9) Taking precautions against real or apparent conflicts of interest on the part of those involved in the inquiry or investigation.

(10) Preparing and maintaining the documentation to substantiate the investigation's findings. This documentation is to be made available to the Director, OSI, who will decide whether that Office will either proceed with its own investigation or will act on the institution's findings.

(11) Taking interim administrative actions, as appropriate, to protect Federal funds and insure that the purposes of the Federal financial assistance are carried out.

(12) Keeping the OSI apprised of any developments during the course of the investigation which disclose facts that may affect current or potential Department of Health and Human Services funding for the individual(s) under investigation or that the PHS needs to know to ensure appropriate use of Federal funds and otherwise protect the public interest.

(13) Undertaking diligent efforts, as appropriate, to restore the reputations of persons alleged to have engaged in misconduct when allegations are not confirmed, and also undertaking diligent efforts to protect the positions and reputations of those persons who, in good faith, make allegations.

(14) Imposing appropriate sanctions on individuals when the allegation of misconduct has been substantiated.

(15) Notifying the OSI of the final outcome of the investigation.

APPENDIX D

45 CODE OF FEDERAL REGULATIONS, SUBTITLE A

§ 46.107 IRB Membership

(a) Each IRB shall have at least five members, with varying backgrounds to promote complete and adequate review of research activities commonly conducted by the institution. The IRB shall be sufficiently qualified through the experience and expertise of its members, and the diversity of the members' backgrounds including consideration of the racial and cultural backgrounds of members and sensitivity to such issues as community attitudes, to promote respect for its advice and counsel in safeguarding the rights and welfare of human subjects. In addition to possessing the professional competence necessary to review specific research activities, the IRB shall be able to ascertain the acceptability of proposed research in terms of institutional commitments and regulations, applicable law, and standards of professional conduct and practice. The IRB shall therefore include persons knowledgeable in these areas. If an IRB regularly reviews research that involves a vulnerable category of subjects, including but not limited to subjects covered by other subparts of this part, the IRB shall include one or more individuals who are primarily concerned with the welfare of these subjects.

(b) No IRB may consist entirely of men or entirely of women, or entirely of members of one profession.

(c) Each IRB shall include at least one member whose primary concerns are in nonscientific areas; for example: lawyers, ethicists, members of the clergy.

(d) Each IRB shall include at least one member who is not otherwise affiliated with the institution and who is not part of the immediate family of a person who is affiliated with the institution.

(e) No IRB may have a member participating in the IRB's initial or continuing review of any project in which the member has a conflicting interest, except to provide information requested by the IRB.

(f) An IRB may, in its discretion, invite individuals with competence in special areas to assist in the review of complex issues which require expertise beyond or in addition to that available on the IRB. These individuals may not vote with the IRB.

§ 46.108 IRB Functions and Operations

In order to fulfill the requirements of these regulations each IRB shall:

(a) Follow written procedures as provided in § 46.103(b)(4).

(b) Except when an expedited review procedure is used (see § 46.110), review proposed research at convened meetings at which a majority of the members of the IRB are present, including at least one member whose primary concerns are in nonscientific areas. In order for the research to be approved, it shall receive the approval of a majority of those members present at the meeting.

(c) Be responsible for reporting to the appropriate institutional officials and the Secretary[1] any serious or continuing noncompliance by investigators with the requirements and determinations of the IRB.

[46 FR 8386, Jan. 26, 1981; 46 FR 19195, Mar. 27, 1981]

§ 46.109 IRB Review of Research

(a) An IRB shall review and have authority to approve, require modifications in (to secure approval), or disapprove all research activities covered by these regulations.

(b) An IRB shall require that information given to subjects as part of informed consent is in accordance with § 46.116. The IRB may require that information, in addition to that specifically mentioned in § 46.116, be given to the subjects when in the IRB's judgment the information would meaningfully add to the protection of the rights and welfare of subjects.

(c) An IRB shall require documentation of informed consent or may waive documentation in accordance with § 46.117.

(d) An IRB shall notify investigators and the institution in writing of its decision to approve or disapprove the proposed research activity, or of modifications required to secure IRB approval of the research activity. If the IRB decides to disapprove a research activity, it shall include in its written notification a statement of the reasons for its decision and give the investigator an opportunity to respond in person or in writing.

(e) An IRB shall conduct continuing review of research covered by these regulations at intervals appropriate to the degree of risk, but not less than once per year, and shall have authority to observe or have a third party observe the consent process and the research.

---

[1]Reports should be filed with the Office for Protection from Research Risks, National Institutes of Health, Department of Health and Human Services, Bethesda, Maryland 20205.

§ 46.116 General Requirements for Informed Consent

Except as provided elsewhere in this or other subparts, no investigator may involve a human being as a subject in research covered by these regulations unless the investigator has obtained the legally effective informed consent of the subject or the subject's legally authorized representative. An investigator shall seek such consent only under circumstances that provide the prospective subject or the representative sufficient opportunity to consider whether or not to participate and that minimize the possibility of coercion or undue influence. The information that is given to the subject or the representative shall be in language understandable to the subject or the representative. No informed consent, whether oral or written, may include any exculpatory language through which the subject or the representative is made to waive or appear to waive any of the subject's legal rights, or releases or appears to release the investigator, the sponsor, the institution or its agents from liability for negligence.

(a) Basic elements of informed consent. Except as provided in paragraph (c) or (d) of this section, in seeking informed consent the following information shall be provided to each subject:

(1) A statement that the study involves research, an explanation of the purposes of the research and the expected duration of the subject's participation, a description of the procedures to be followed, and identification of any procedures which are experimental;

(2) A description of any reasonably foreseeable risks or discomforts to the subject;

(3) A description of any benefits to the subject or to others which may reasonably be expected from the research;

(4) A disclosure of appropriate alternative procedures or courses of treatment, if any, that might be advantageous to the subject;

(5) A statement describing the extent, if any, to which confidentiality of records identifying the subject will be maintained;

(6) For research involving more than minimal risk, an explanation as to whether any compensation and an explanation as to whether any medical treatments are available if injury occurs and, if so, what they consist of, or where further information may be obtained;

(7) An explanation of whom to contact for answers to pertinent questions about the research and research subjects' rights, and whom to contact in the event of a research-related injury to the subject; and

(8) A statement that participation is voluntary, refusal to participate will involve no penalty or loss of benefits to which the subject is otherwise entitled, and the subject may discontinue participation at any time without penalty or loss of benefits to which the subject is otherwise entitled.

(b) Additional elements of informed consent. When appropriate, one or more of the following elements of information shall also be provided to each subject:

(1) A statement that the particular treatment or procedure may involve risks to the subject (or to the embryo or fetus, if the subject is or may become pregnant) which are currently unforeseeable;

(2) Anticipated circumstances under which the subject's participation may be terminated by the investigator without regard to the subject's consent;

(3) Any additional costs to the subject that may result from participation in the research;

(4) The consequences of a subject's decision to withdraw from the research and procedures for orderly termination of participation by the subject;

(5) A statement that significant new findings developed during the course of the research which may relate to the subject's willingness to continue participation will be provided to the subject; and

(6) The approximate number of subjects involved in the study.

(c) An IRB may approve a consent procedure which does not include, or which alters, some or all of the elements of informed consent set forth above, or waive the requirement to obtain informed consent provided the IRB finds and documents that:

(1) The research or demonstration project is to be conducted by or subject to the approval of state or local government officials and is designed to study, evaluate, or otherwise examine: (i) Programs under the Social Security Act, or other public benefit or service programs; (ii) procedures for obtaining benefits or services under those programs; (iii) possible changes in or alternatives to those programs or procedures; or (iv) possible changes in methods or levels of payment for benefits or services under those programs; and

(2) The research could not practicably be carried out without the waiver or alteration.

(d) An IRB may approve a consent procedure which does not include, or which alters, some or all of the elements of informed consent set forth above, or waive the requirements to obtain informed consent provided the IRB finds and documents that:

(1) The research involves no more than minimal risk to the subjects;

(2) The waiver or alteration will not adversely affect the rights and welfare of the subjects;

(3) The research could not practicably be carried out without the waiver or alteration; and

(4) Whenever appropriate, the subjects will be provided with additional pertinent information after participation.

(e) The informed consent requirements in these regulations are not intended to preempt any applicable Federal, state, or local laws which require additional information to be disclosed in order for informed consent to be legally effective.

(f) Nothing in these regulations is intended to limit the authority of a physician to provide emergency medical care, to the extent the physician is permitted to do so under applicable Federal, state, or local law.

APPENDIX E

9 CODE OF FEDERAL REGULATIONS, CHAPTER 1

§ 2.31 Institutional Animal Care and Use Committee (IACUC)

(a) The Chief Executive Officer of the research facility shall appoint an Institutional Animal Care and Use Committee (IACUC), qualified through the experience and expertise of its members to assess the research facility's animal program, facilities, and procedures. Except as specifically authorized by law or these regulations, nothing in this part shall be deemed to permit the Committee or IACUC to prescribe methods or set standards for the design, performance, or conduct of actual research or experimentation by a research facility.

(b) IACUC Membership. (1) The members of each Committee shall be appointed by the Chief Executive Officer of the research facility;

(2) The Committee shall be composed of a Chairman and at least two additional members;

(3) Of the members of the Committee:

(i) At least one shall be a Doctor of Veterinary Medicine, with training or experience in laboratory animal science and medicine, who has direct or delegated program responsibility for activities involving animals at the research facility;

(ii) At least one shall not be affiliated in any way with the facility other than as a member of the Committee, and shall not be a member of the immediate family of a person who is affiliated with the facility. The Secretary intends that such person will provide representation for general community interests in the proper care and treatment of animals;

(4) If the Committee consists of more than three members, not more than three members shall be from the same administrative unit of the facility.

(c) IACUC Functions. With respect to activities involving animals, the IACUC, as an agent of the research facility, shall:

(1) Review, at least once every six months, the research facility's program for humane care and use of animals, using title 9, chapter I, subchapter A—Animal Welfare, as a basis for evaluation;

(2) Inspect, at least once every six months, all of the research facility's animal facilities, including animal study areas, using title 9, chapter I, subchapter A—Animal Welfare, as a basis for evaluation; *Provided, however,* That animal areas containing free-living wild animals in their natural habitat need not be included in such inspection;

(3) Prepare reports of its evaluations conducted as required by paragraphs (c) (1) and (2) of this section, and submit the reports to the Institutional Official of the research facility; *Provided, however,* That the IACUC may determine the best means of conducting evaluations of the research facility's programs and facilities; and *Provided, further,* That no Committee member wishing to participate in any evaluation conducted under this subpart may be excluded. The IACUC may use subcommittees composed of at least two Committee members and may invite *ad hoc* consultants to assist in conducting the evaluations, however, the IACUC remains responsible for the evaluations and reports as required by the Act and regulations. The reports shall be reviewed and signed by a majority of the IACUC members and must include any minority views. The reports shall be updated at least once every six months upon completion of the required semiannual evaluations and shall be maintained by the research facility and made available to APHIS and to officials of funding Federal agencies for inspection and copying upon request. The reports must contain a description of the nature and extent of the research facility's adherence to this subchapter, must identify specifically any departures from the provisions of title 9, chapter I, subchapter A—Animal Welfare, and must state the reasons for each departure. The reports must distinguish significant deficiencies from minor deficiencies. A significant deficiency is one which, with reference to Subchapter A, and, in the judgment of the IACUC and the Institutional Official, is or may be a threat to the health or safety of the animals. If program or facility deficiencies are noted, the reports must contain a reasonable and specific plan and schedule with dates for correcting each deficiency. Any failure to adhere to the plan and schedule that results in a significant deficiency remaining uncorrected shall be reported in writing within 15 business days by the IACUC, through the Institutional Official, to APHIS and any Federal agency funding that activity;

(4) Review, and, if warranted, investigate concerns involving the care and use of animals at the research facility resulting from public complaints received and from reports of noncompliance received from laboratory or research facility personnel or employees;

(5) Make recommendations to the Institutional Official regarding any aspect of the research facility's animal program, facilities, or personnel training;

(6) Review and approve, require modifications in (to secure approval), or withhold approval of those components of proposed activities related to the care and use of animals, as specified in paragraph (d) of this section;

(7) Review and approve, require modifications in (to secure approval), or withhold approval of proposed significant changes regarding the care and use of animals in ongoing activities; and

(8) Be authorized to suspend an activity involving animals in accordance with the specifications set forth in paragraph (d)(6) of this section.

(d) IACUC review of activities involving animals. (1) In order to approve proposed activities or proposed significant changes in ongoing activities, the IACUC shall conduct a review of those components of the activities related to the care and use of animals and determine that the proposed activities are in accordance with this subchapter unless acceptable justification for a departure is presented in writing; *Provided, however,* That field studies as defined in part 1 of this subchapter are exempt from this requirement. Further, the IACUC shall determine that the proposed activities or significant changes in ongoing activities meet the following requirements:

(i) Procedures involving animals will avoid or minimize discomfort, distress, and pain to the animals;

(ii) The principal investigator has considered alternatives to procedures that may cause more than momentary or slight pain or distress to the animals, and has provided a written narrative description of the methods and sources, e.g., the Animal Welfare Information Center, used to determine that alternatives were not available;

(iii) The principal investigator has provided written assurance that the activities do not unnecessarily duplicate previous experiments;

(iv) Procedures that may cause more than momentary or slight pain or distress to the animals will:

(A) Be performed with appropriate sedatives, analgesics or anesthetics, unless withholding such agents is justified for scientific reasons, in writing, by the principal investigator and will continue for only the necessary period of time;

(B) Involve, in their planning, consultation with the attending veterinarian or his or her designee;

(C) Not include the use of paralytics without anesthesia;

(v) Animals that would otherwise experience severe or chronic pain or distress that cannot be relieved will be painlessly euthanized at the end of the procedure or, if appropriate, during the procedure;

(vi) The animals' living conditions will be appropriate for their species in accordance with part 3 of this subchapter, and contribute to their health and comfort. The housing, feeding, and nonmedical care of the animals will be directed by the attending veterinarian or other scientist trained and experienced in the proper care, handling, and use of the species being maintained or studied;

(vii) Medical care for animals will be available and provided as necessary by a qualified veterinarian;

(viii) Personnel conducting procedures on the species being maintained or studied will be appropriately qualified and trained in those procedures;

(ix) Activities that involve surgery include appropriate provision for pre-operative and post-operative care of the animals in accordance with established veterinary medical and nursing practices. All survival surgery will be performed using aseptic procedures, including surgical gloves, masks, sterile instruments, and aseptic techniques. Major operative procedures on non-rodents will be conducted only in facilities intended for that purpose which shall be operated and maintained under aseptic conditions. Non-major operative procedures and all surgery on rodents do not require a dedicated facility, but must be performed using aseptic procedures. Operative procedures conducted at field sites need not be performed in dedicated facilities, but must be performed using aseptic procedures;

(x) No animal will be used in more than one major operative procedure from which it is allowed to recover, unless:

(A) Justified for scientific reasons by the principal investigator, in writing;

(B) Required as routine veterinary procedure or to protect the health or well-being of the animal as determined by the attending veterinarian; or

(C) In other special circumstances as determined by the Administrator on an individual basis. Written requests and supporting data should be sent to the Administrator, APHIS, USDA, 6505 Belcrest Road, Room 268, Hyattsville, MD 20782;

(xi) Methods of euthanasia used must be in accordance with the definition of the term set forth in 9 CFR part 1, § 1.1 of this subchapter, unless a deviation is justified for scientific reasons, in writing, by the investigator.

(2) Prior to IACUC review, each member of the Committee shall be provided with a list of proposed activities to be reviewed. Written descriptions of all proposed activities that involve the care and use of animals shall be available to all IACUC members, and any member of the IACUC may obtain, upon request, full Committee review of those activities. If full Committee review is not requested, at least one member of the IACUC, designated by the chairman and qualified to conduct the review, shall review those activities, and shall have the authority to approve, require modifications in (to secure approval), or request full Committee review of any of those activities. If full Committee review is requested for a proposed activity, approval of that activity may be granted only after review, at a convened meeting of a quorum of the IACUC, and with the approval vote of a majority of the quorum present. No member may participate in the IACUC review or approval of an activity in which that member has a conflicting interest (e.g., is personally involved in the activity), except to provide information requested by the IACUC, nor may a member who has a conflicting interest contribute to the constitution of a quorum;

(3) The IACUC may invite consultants to assist in the review of complex issues arising out of its review of proposed activities. Consultants may not approve or withhold approval of an activity, and may not vote with the IACUC unless they are also members of the IACUC;

(4) The IACUC shall notify principal investigators and the research facility in writing of its decision to approve or withhold approval of those activities related to the care and use of animals, or of modifications required to secure IACUC approval. If the IACUC decides to withhold approval of an activity, it shall include in its written notification a statement of the reasons for its decision

and give the principal investigator an oportunity to respond in person or in writing. The IACUC may reconsider its decision, with documentation in Committee minutes, in light of the information provided by the principal investigator;

(5) The IACUC shall conduct continuing reviews of activities covered by this subchapter at appropriate intervals as determined by the IACUC, but not less than annually;

(6) The IACUC may suspend an activity that it previously approved if it determines that the activity is not being conducted in accordance with the description of that activity provided by the principal investigator and approved by the Committee. The IACUC may suspend an activity only after review of the matter at a convened meeting of a quorum of the IACUC and with the suspension vote of a majority of the quorum present;

(7) If the IACUC suspends an activity involving animals, the Institutional Official, in consultation with the IACUC, shall review the reasons for suspension, take appropriate corrective action, and report that action with a full explanation to APHIS and any Federal agency funding that activity; and

(8) Proposed activities and proposed significant changes in ongoing activities that have been approved by the IACUC may be subject to further appropriate review and approval by officials of the research facility. However, those officials may not approve an activity involving the care and use of animals if it has not been approved by the IACUC.

(e) A proposal to conduct an activity involving animals, or to make a significant change in an ongoing activity involving animals, must contain the following:

(1) Identification of the species and the approximate number of animals to be used;

(2) A rationale for involving animals and for the appropriateness of the species and numbers of animals to be used;

(3) A complete description of the proposed use of the animals;

(4) A description of procedures designed to assure that discomfort and pain to animals will be limited to that which is unavoidable for the conduct of scientifically valuable research, including provision for the use of analgesic, anesthetic, and tranquilizing drugs where indicated and appropriate to minimize discomfort and pain to animals; and

(5) A description of any euthanasia method to be used.

REFERENCES

American Association for Laboratory Animal Science. 1963. *The guide for the care and use of laboratory animals.* New York: American Association for Laboratory Animal Science.

American Law Institute. 1958. *Model penal code.* Philadelphia: American Law Institute.

Arluke, A.B. 1988. Sacrificial symbolism in animal experimentation: Object or pet? *Anthrozoos* 2:98-117.

Association of American Medical Colleges. 1982. *The maintenance of high ethical standards in the conduct of research.* Washington, D.C.: Association of American Medical Colleges.

Boruch, R.F. 1975. *Is a promise of confidentiality necessary? Sufficient? A review and bibliography* (Research Report NIE-11/11X). Evanston, Ill.: Northwestern University, Evaluation Research Program.

Brackbill, Y., and A.E. Hellegers. 1980. Hasting Center Report, *Ethics and editors* 10, no. 2:20-22.

Callahan, D. 1982. Should there be an academic code of ethics? *Journal of Higher Education* 53:335-44.

Committee on Care and Use of Laboratory Animals of the Institute of Laboratory Animal Resources. 1985. *Guide for the care and use of laboratory animals.* 5th ed. Bethesda, MD: National Institutes of Health. NIH Publication No. 86-23.

Dill, C.A., E.R. Gliden, P.C. Hill, and L.L. Hanselka. 1982. Federal human subject regulations: A methodological artifact? *Personality and Social Psychology Bulletin* 8:417-25.

Final Rule. Responsibilities of Awardee and Applicant Institutions for Dealing with and Reporting Possible Misconduct in Science. Public Health Service Act. September 1989, vol. 18, no. 30. Final Rule also published in the *Federal Register* 54, no. 151 (8 August 1989): 32446-51.

42 *Code of Federal Regulations* (C.F.R.), chapter 1, part 50. DHHS Regulations: October 1, 1991.

45 *Code of Federal Regulations* (C.F.R.), subtitle A, part 46. DHHS Regulations: October 1, 1990.

Gardner, G.T. 1978. Effects of federal human subjects regulations on data obtained in environmental stressor research. *Journal of Personality and Social Psychology* 36:628-34.

Gorman, C. 1994. Medicine: Let's not be too hasty. *Time*, 19 September, 71.

Grassian, V. 1992. *Moral reasoning: Ethical theory and some contemporary moral problems.* 2d ed. Englewood Cliffs: Prentice Hall.

Huff, D. 1954. *How to lie with statistics.* New York: Norton.

Kaemingk, K.L., and L. Sechrest. 1990. AIDS research and policy decisions. *Evaluation and Program Planning* 13:1-7.

Kaufman, S.R., and B. Todd, eds. 1989. *Perspectives on animal research.* Vol. 1. New York: Medical Research Modernization Committee.

Kerlinger, F.N. 1966. *Foundations of behavioral research: Educational and psychological inquiry.* New York: Holt, Rinehart and Winston.

Kimmel, A.J. 1988. *Ethics and values in applied social research.* Newbury Park, Calif.: Sage.

Kretchmar, R.S. 1993. Philosophy of ethics. *Quest* 45:3-12.

Kroll, W. 1993. Ethical issues in human research. *Quest* 45:32-44.

Levine, R.J. 1981. *Ethics and regulation of clinical research.* Baltimore: Urban & Schwarzenberg.

Massey, B.H. 1973. Principles of problem solving. In *Research methods in health, physical education, and recreation,* 3d ed., ed. A.W. Hubbard. Washington, D.C.: AAHPERD.

Matt, K.S. 1993. Ethical issues in animal research. *Quest* 45:45-51.

National Commission for Protection of Human Subjects of Biomedical and Behavioral Research. 1976. *Research involving prisoners: Report and recommendations.* (DHEW Publication No. (OS) 76-131). Washington, D.C.: Government Printing Office. The appendix to this report is DHEW Publication No. (OS) 76-132.

———. 1977. *Research involving children: Report and recommendations.* (DHEW Publication No. (OS) 77-0004). Washington, D.C.: Government Printing Office. The appendix to this report is DHEW Publication No. (OS) 77-005.

———. 1978. *Institutional review boards: Report and recommendations.* (DHEW Publication No. (OS) 78-0008). Washington, D.C.: Government Printing Office. The appendix to this Report is DHEW Publication no. (OS) 77-009.

National Institutes of Health. 1990. *Guidelines for the conduct of research at the National Institutes of Health* (21 March 1990). Bethesda, Md.: National Institutes of Health.

National Research Act. U.S. Public Law 93-348 (1974).

9 *Code of Federal Regulations* (C.F.R.), chapter 1, part 2. USDA Regulations: January 1, 1990.

Nuremberg Code. 1947. Reprinted from *Trials of war criminals before the Nuremberg military tribunals under Control Council law,* vol. 2, no. 10 (Washington, D.C.: Government Printing Office, 1949), 181-82.

Pring, R. 1984. Confidentiality and the right to know. In *The politics and ethics of education,* ed. C. Adelman. London: Croom Helm.
Public Health Service. [1992?] *Institutional animal care and use committee guidebook.* Washington, D.C.: Public Health Service. (NIH Publication No. 92-3415)
Reagan, C.E. 1971. *Ethics for scientific researchers.* 2d ed. Springfield, Ill.: Charles C Thomas.
Reece, R.D., and H.A. Siegal. 1986. *Studying people: A primer in the ethics of social research.* Macon, Ga.: Mercer University Press.
Resnick, J.H., and T. Schwartz. 1973. Ethical standards as an independent variable in psychological research. *American Psychologist* 39:863-76.
Roethlisberger, F.J., and W.J. Dickson. 1939. *Management and the worker.* Cambridge: Harvard University Press.
Rosenthal, R., and L. Jacobson. 1968. *Pygmalion in the classroom.* New York: Holt, Rinehart and Winston.
Russell, P.S. 1991. Setting sanctions for misconduct. Paper presented at symposium, Scientific Integrity: Major Issues Faced by Research Institutions, February 1991, at Harvard Medical School.
Shore, E.G. 1991. Analysis of a multi-institutional series of completed cases. Paper presented at symposium, Scientific Integrity: Major Issues Faced by Research Institutions, February 1991, at Harvard Medical School.
Singer, E. 1978. Informed consent: Consequences for response rate and response quality in social surveys. *American Sociological Review* 43:144-62.
World Medical Association Declaration of Helsinki: Recommendations guiding medical doctors in biomedical research involving human subjects, adopted by the 18th World Medical Assembly, Helsinki, 1964, and as revised by the 29th World Medical Assembly, Tokyo, 1975.
Wrather, J. 1987. Scientists and lawyers look at fraud in science. *Science* 238:813-814.
Zelaznik, H.N. 1993. Ethical issues in conducting and reporting research: A reaction to Kroll, Matt, and Safrit. *Quest* 45:62-68.

Additional Resources

American Psychological Association. 1953. *Ethical standards of psychologists.* Washington, D.C.: American Psychological Association.
———. 1974. *Standards for educational and psychological tests.* Washington, D.C.: American Psychological Association.
———. 1977. *Standards for providers of psychological services.* Rev. ed. Washington, D.C.: American Psychological Association.
———. 1981a. *Ethical principles of psychologists.* Washington, D.C.: American Psychological Association.
———. 1981b. Specialty guidelines for the delivery of services by industrial/organizational psychologists. *American Psychologist* 36:664-69.

———. 1982. *Ethical principles in the conduct of research with human participants.* Washington, D.C.: American Psychological Association.

American Sociological Association. 1971. *Code of ethics.* Washington, D.C.: American Sociological Association.

Animal Welfare Act of 1970 [with 1985 amendments]. *U.S. Code.* Vol. 7, secs. 2131 *et seq.* (1985).

Association of American Medical Colleges. 1990. *Framework for institutional policies and procedures to deal with misconduct in research.* Rev. ed. Washington, D.C.: Division of Biomedical Research.

Bok, S. 1978. Freedom and risk. *Proceedings of the American Academy of Arts and Sciences* 107:115-28.

Byrne, P., ed. 1990. *Ethics and law in health care and research.* New York: Wiley.

Callahan, D., and H.T. Englehardt, eds. 1981. *Roots of ethics: Science, religion, and values.* New York: Plenum Press.

Committee on the Conduct of Science, National Academy of Sciences. 1989. *On being a scientist.* Washington, D.C.: National Academy Press.

Davis, B.D. 1990. How far should Big Brother's hand reach? *ASM News* 56: 643-46.

Editorial Committee of Institutional Administrators and Laboratory Animal Specialists for the Henry M. Jackson Foundation Advancement of Military Medicine. 1988. *Institutional administrator's manual for laboratory animal care and use.* Bethesda, MD: National Institutes of Health. NIH Publication No. 88-2959.

Food and Drug Administration. 1990. ''Fraud, material false statements, bribery and illegal gratuities; proposed policy.'' *Federal Register* 55, no. 246 (21 December): 52323-25.

Hallum, J.V., and S.W. Hadley. 1990. OSI: Why, what, and how? *ASM News* 56:647-51.

Harvard University, Faculty of Medicine. 1988. Guidelines for investigators in scientific research. Adopted 16 February 1988.

Harvard University, Faculty of Medicine. 1989. ''Principles and procedures for dealing with allegations of faculty misconduct.'' Adopted 14 December 1989.

Hunter, E., J. Mayotte, and D. Junter, eds. 1982. *Black, white, and gray: Issues in professional ethics and law.* San Diego: Hunter.

Institute of Medicine. 1989. *Report of a study: The responsible conduct of research in the health sciences.* Washington, D.C.: National Academy Press.

Katz, J. 1927. *Experimentation with human beings.* New York: Russell Sage Foundation.

Levy, C. 1983. *The human body and the law: Legal and ethical considerations in human experimentation.* 2d ed. New York: Oceana.

Mishkin, B. 1988. Responding to scientific misconduct due process and prevention. *Journal of the American Medical Association* 260:1932-36.

Morley, D. 1978. *The sensitive scientist.* London: SCM Press.

National Science Foundation regulations concerning allegations of misconduct in science, engineering, or education. NSF letter, 28 September 1990.

Newkirk, I. 1990. *Save the animals: 101 easy things you can do*. New York: Warner Books.

President's Commission for the Study of Ethical Problems in Medicine and Biomedical and Behavioral Research. 1983, March. *Implementing human research regulations: Second biennial report on the adequacy and uniformity of federal rules and policies, and of their implementation, for the protection of human subjects*. Washington, D.C.: President's Commission.

Public Health Service. 1990. "Policies and procedures for dealing with possible misconduct in extramural research." Dated 25 July 1990. Washington D.C.: Public Health Service.

Sigma Xi. 1986. *Honor in science*. Research Triangle Park, NC: Sigma Xi Scientific Research Society.

INDEX

ABOUT THE AUTHOR

John N. Drowatzky, JD, EdD, is professor emeritus of exercise science in the Department of Health Promotion and Human Performance at the University of Toledo, where he has served on the faculty since 1965.

As both a lawyer and physical educator, Dr. Drowatzky has long been aware of the problems associated with professional ethics and the potential for abuse in the research process. This awareness, combined with a desire to increase public confidence in research and the people who undertake it, prompted Dr. Drowatzky to study the relationship between general ethical issues and specific research practices. His articles on the subject have appeared in such prestigious journals as *Clinical Kinesiology* and *Quest*. In addition, Dr. Drowatzky has chaired both the Ethics Committee of the American Academy of Physical Education and Kinesiology and the Ethics Committee for the Research Consortium of the American Alliance for Health, Physical Education, Recreation and Dance (AAHPERD).

Dr. Drowatzky received his bachelor of science in physical education and social science from the University of Kansas. He went on to complete a master of science in physical education administration and a doctorate in scientific foundations of physical education at the University of Oregon. In 1979 Dr. Drowatzky earned a Juris Doctor from the University of Toledo and passed the Ohio Bar exam.

Dr. Drowatzky is the author of nine books and more than 50 scholarly articles on a variety of topics in physical education and the law. He is a fellow in the Research Consortium of AAHPERD and was a member of the International Society of Sports Psychology, the North American Society for the Psychology of Sport and Physical Activity, and the American, Ohio, and Toledo Bar Associations.

Dr. Drowatzky resides in suburban Toledo, Ohio, with his wife Linnea. His favorite leisure activities include playing tennis and golf, reading, and traveling.